Rehabilitating Damaged Ecosystems

Volume I

Editor

John Cairns, Jr.

University Distinguished Professor
Department of Biology
and
Director
University Center for Environmental and Hazardous Materials Studies
Virginia Polytechnic Institute and State University
Blacksburg, Virginia

CRC Press, Inc.
Boca Raton, Florida

Library of Congress Cataloging-in-Publication Data

Rehabilitating damaged ecosystems.

  Includes bibliographies and indexes.
  1. Restoration ecology--Case studies. I. Cairns,
John, 1923- .
QH541.15.R45R47 1988   631.6'4   87-21842
ISBN 0-8493-4391-7 (v. 1)
ISBN 0-8493-4392-5 (v. 2)

Direct all inquiries to CRC Press, Inc., 2000 Corporate Blvd., N.W., Boca Raton, Florida, 33431.

© 1988 by CRC Press, Inc.

International Standard Book Number 0-8493-4391-7 (Volume I)
International Standard Book Number 0-8493-4392-5 (Volume II)

Library of Congress Card Number 87-21842
Printed in the United States

DEDICATION

To the Memory of René Dubos

This book is dedicated to René Dubos, scientist, humanist, and environmentalist, who believed that:

> The earth is to be seen neither as an ecosystem to be preserved unchanged nor as a quarry to be exploited for selfish and short-range economic reasons but as a garden to be cultivated for the development of its own potentialities of the human adventure.*

* Dubos, R., Symbiosis between the earth and mankind, *Science,* 193, 462, 1976.

PREFACE

Readers of this volume may well conclude that a different sequencing of chapters would be more desirable than the one used. A number of alternatives were considered before arriving at the final sequence. When I designed the content of the book several years ago, I had an enormous list of topics and prospective authors. Many of these had to be eliminated because the size of the book would have been unacceptable, and a significant number of authors who agreed to contribute nearly 2 years ago were unable to meet the deadline due to personal problems, etc. For example, a geologist skilled in establishing soil and rock profiles that would markedly decrease groundwater contamination had so many pressures on his time due to the multiplicity of groundwater problems that he was unable to meet the agreed upon deadline even though this was twice extended. Other invited contributors withdrew at various stages but, without exception, well after the time for replacing them had passed. As a consequence, the original sequencing was inappropriate. Nevertheless, having read the contributed chapters several times and after comparing them with the few other volumes in this field, the aggregate still represents a unique contribution that should advance our understanding of the field.

The sequencing ultimately chosen was to place William R. Jordan's chapter at the beginning to relate the field of restoration ecology to ecological research as a whole. This was followed by a series of case histories that could be taken in any order (Gore and Bryant; Grunwald, Iverson, and Szafoni; Smith; Thompson; Willard; Zedler; Daniels and Zipper; Bruns). The number of case histories could easily have been increased an order of magnitude without exhausting the possibilities in this rapidly expanding field. Those chosen were meant to provide illustrative cases of successful approaches used in quite different ecosystems. By using the references in the chapters collectively, one can gain access to a very substantial number of other case histories. In addition, one of the earliest series of case histories is in the volume *Recovery and Restoration of Damaged Ecosystems*.[1] Following the case histories of rehabilitation and restoration following anthropogenic stress, there is the article by Franklin, Frenzen, and Swanson ("Re-Creation of Ecosystems at Mount St. Helens: Contrasts in Artificial and Natural Approaches") which provides a linkage with some of the topics to follow. The chapter by Inouye on variation is important because variability is one of the important characteristics of natural systems that is often poorly documented and equally important. If ignored, variability might result in a rehabilitation process being unacceptable to a regulatory agency if a downtrend due to natural variability is not distinguished from a failure in restoration ecology. This is followed by a chapter by Louda on "Insect Pests and Plant Stress as Considerations for Revegetation of Disturbed Ecosystems." The point of this chapter is that species representing a relatively small component of the biomass may nevertheless be important regulators of community structure. Failure to include them in the restoration process may result in a community or ecosystem quite different from the one intended. This chapter merely illustrates the importance of species that might not be viewed as dominants but are nevertheless important in determining density relationships of other species, etc. This is an illustrative situation calling attention to examination of how communities and ecosystems work rather than an attempt to call attention to a specific plant/animal relationship. The paper by Bradshaw, "Alternative Endpoints for Reclamation," gives an overview from one of the major figures in restoration ecology with a vast experience in this field. The article by Plafkin provides the perspective of a person employed by a federal regulatory agency, although it does not represent the view of the agency but rather of an individual. The article by Klose, Rich, and Schneider touches on the importance of political and social factors in a restoration or clean-up project. Following this is an article by Maguire on decision analysis which is, in my opinion, the best approach to resolving conflict in situations where the outcome is often uncertain. Although restoration ecology can improve damaged ecosystems, the precise outcome of restoration activities cannot be predicted with a high degree of certainty and this, coupled with the fact that the final product and end-points may be in dispute as a result of varying demands from citizens, etc., requires that a systematic and orderly way of resolving these conflicts and issues be implemented at the outset. The chapter by Ehrlich is included because the ultimate problem in restoration ecology will occur if the political system fails and there is even a limited nuclear exchange. Although there is general agreement that the biological consequences of even a modest nuclear winter will be extremely deleterious, the prospects for rehabilitation are highly uncertain. Nevertheless, I persuaded Ehrlich that even documenting the degree of uncertainty about our ability to restore the earth to even a modest semblance of its prenuclear winter condition is a sufficiently sobering exercise to perhaps enhance the chances for restraint on the part of the nuclear powers. Janzen's article shows a beautiful balance of sound ecology and practicality in restoring a once vast ecosystem now almost entirely gone. This manuscript arrived in June 1986 and, at the National Forum on BioDiversity cosponsored by the National Academy of Sciences and the Smithsonian Institution held from September 21 to 24, 1986, I learned from Dan that he is now well on his way toward the second million dollars of the amount discussed in the article. What began as a vision may yet become a reality. Hugo Ferchau's chapter discusses rehabilitation under difficult and unusual ecological circumstances.

Future management options will be dramatically affected by the degree of skill used in providing research opportunities in ecosystems that industry is charged with rehabilitating or in derelict ecosystems where some level of government must accept this responsibility since it is legally impossible to assign this to an individual or a private institution. The young and rapidly developing field of restoration ecology has evolved to a sufficient degree to enable cost-effective and ecologically interesting restoration of ecosystems damaged by mining and other anthropogenic activities.[2] However, it is abundantly clear that regulatory measures may impede the generation of an adequate scientific base for further development of the field.[3,4] It is important for research investigators, regulators, industries, and other organizations charged with rehabilitation to realize that they have a common interest but that all will have to adjust to the needs of the others in order to achieve the desired results. These desired results should be (1) a cost-effective means of fulfilling the responsibility of rehabilitating the damaged ecosystems so that those charged with this responsibility meet the legal requirements for doing so; (2) the regulators must be able to demonstrate that they have fulfilled their responsibility to the public and to their superiors in the government at whatever level of government they represent; and (3) the research investigators must derive sufficient new information to justify publication in a peer-reviewed scholarly journal. It is apparent from *Wetlands and Water Management on Mined Lands*,[3] that, while we might do much better in this regard, a heartening degree of success has already been achieved.

The number of management options that can be implemented is a direct function of the size and reliability of the information base available. The scientific information base is not likely to increase at an acceptable rate unless a substantial amount of the money comes from nontraditional sources. Ecosystem rehabilitation is a major undertaking and will require a level of funding not likely to become available from the National Science Foundation, the National Institutes of Health, or many of the other traditional sources of research funding because of the inability to meet present demands, let alone satisfy new ones. On the other hand, research investigators are not likely to solicit funds from nontraditional sources unless they can get the most critical of their needs met by them, namely (1) the ability to carry out experiments that can be published in scholarly peer-reviewed journals and (2) recognition by their peers that these efforts represent good science. Without this research effort, regulators will be enforcing requirements which in the present state of knowledge have a highly uncertain ecological outcome and may, in fact, frequently fail totally. In the absence of an adequate information base, industry is faced with a higher degree of uncertainty regarding rehabilitation costs and probable success than most industrial managers consider desirable. Therefore, both regulators and industrial groups can increase the certainty of the successful outcome of a rehabilitation effort and simultaneously satisfy the needs of research investigators if there is skillful communication about the needs that each group needs to have satisfied. This will not be an easy task because the basis of a satisfactory working relationship is mutual trust, and the three groups just mentioned have often been in contention rather than working effectively together. Nevertheless, industrial and university ties are increasing and, while there are some dangers to this process, there are also many opportunities. There has been a less significant development in interactions between regulators and universities, but even here there are some grounds for optimism. Many of the chapters in this book will show the "state of the art" methodology that will markedly influence present management options for rehabilitation of damaged ecosystems. The degree to which this relationship just described flourishes will, in large part, determine the future management options.

John Cairns, Jr.

REFERENCES

1. **Cairns, J., Jr., Dickson, K. L., and Herricks, E. E., Eds.,** *Recovery and Restoration of Damaged Ecosystems,* University Press of Virginia, Charlottesville, 1977.
2. **Brooks, R. P., Samuel, D. E., and Hill, J.B., Eds.,** *Wetlands and Water Management of Mined Lands,* The Pennsylvania State University, University Park, Pa., 1985.
3. **Wyngaard, G. A.,** Ethical and ecological concerns in land reclamation policy: an analysis of the Surface Mining Control and Reclamation Act of 1977, in *Wetlands and Water Management of Mined Lands,* Brooks, R. P., Samuel, D. E., and Hill, J. B., Eds., The Pennsylvania State University, University Park, Pa., 1985, 75.
4. **Cairns, J., Jr. and Pratt, J. R.,** Conference summary —aquatic environments on mined lands: progress and future needs, in *Wetlands and Water Management of Mined Lands,* Brooks, R. P., Samuel, D. E., and Hill, J. B., Eds., The Pennsylvania State University, University Park, Pa., 1985, 1.

ACKNOWLEDGMENTS

I am deeply indebted to Darla Donald, Editorial Assistant in the University Center for Environmental Studies, for her efforts in assembling this volume. Betty Higginbotham typed many pages to avoid returning manuscripts to authors. The following persons served as reviewers, and I gratefully acknowledge their insights and assistance in finalizing the volume.

David J. Allen
Robert Brooks
Ralph E. Clark, III
W. David Conn
Kenneth L. Dickson
Ruth Eblen
William R. Eblen
Robert Grant
Mary Ellen Harte
Janice B. Hill
Carter Johnson
Charles A. Kennedy

William Mitsch
Stuart E. Neff
Robert A. Paterson
Duncan M. Porter
James R. Pratt
A. C. Samli
Terry Sharik
George M. Simmons, Jr.
David W. Smith
Arthur Snoke
Michael Soule
William E. Winner

THE EDITOR

Dr. John Cairns, Jr. is University Distinguished Professor in the Department of Biology and Director of the University Center for Environmental Studies at Virginia Polytechnic Institute and State University, Blacksburg. He received an A.B. from Swarthmore College and his M.S. and Ph.D. from the University of Pennsylvania, and completed a postdoctoral course in isotope methodology at Hahnemann Medical College, Philadelphia. He was Curator of Limnology at the Academy of Natural Sciences of Philadelphia for 18 years, and has taught at various universities and field stations.

Among his awards are the Presidential Commendation in 1971; the Charles B. Dudley Award in 1978 for excellence in publications from the American Society for Testing and Materials; the Superior Achievement Award, U.S. Environmental Protection Agency in 1980; the Founder's Award of the Society for Environmental Toxicology and Chemistry in 1981; the Icko Iben Award for Interdisciplinary Research from the American Water Resources Association in 1984; and the B. Y. Morrison Medal for Outstanding Accomplishments in the Environmental Sciences from the Research Service of the U. S. Department of Agriculture in 1984. A member of many professional societies, he is a member of the Water Science and Technology Board of the National Research Council. Dr. Cairns has been consultant and researcher for the government and private industries, and has served on numerous scientific committees.

His most recent publications are *Multispecies Toxicity Testing* (Pergamon Press, 1985), *Ecoaccidents* (Plenum Press, 1985), *Environmental Regeneration II: Managing Water Resources* (with Ruth Patrick; Praeger Scientific Publishers, 1986), and *Community Toxicity Testing* (American Society for Testing and Materials, 1986).

CONTRIBUTORS

Diedrich Bruns
Assistant Professor
Institut für Landschaftsplanung
Universität Stuttgart
Stuttgart, West Germany

Franklin L. Bryant
Research Assistant
Faculty of Biological Science
University of Tulsa
Tulsa, Oklahoma

John Cairns, Jr.
University Distinguished Professor
Department of Biology
and
Director
University Center for Environmental and
  Hazardous Materials Studies
Virginia Polytechnic Institute and State
  University
Blacksburg, Virginia

W. Lee Daniels
Assistant Professor
Soil and Environmental Sciences
  Section
Department of Agronomy
Virginia Polytechnic Institute and State
  University
Blacksburg, Virginia

James A. Gore
Associate Professor
Faculty of Biological Science
University of Tulsa
Tulsa, Oklahoma

Claus Grunwald
Plant Physiologist and Professor
Section of Botany and Plant Pathology
Illinois Natural History Survey
Champaign, Illinois

Louis R. Iverson
Terrestrial Plant Ecologist
Section of Botany and Plant Pathology
Illinois Natural History Survey
Champaign, Illinois

William R. Jordan, III
Arboretum and Center for Restoration
  Ecology
University of Wisconsin-Madison
Madison, Wisconsin

Michael A. Smith
Bostock Hill and Rigby, Ltd.
Birmingham, England

Diane B. Szafoni
Research Associate
Section of Botany and Plant Pathology
Illinois Natural History Survey
Champaign, Illinois

Carol S. Thompson
President
Restoration Resources, Inc.
Indianapolis, Indiana

Daniel E. Willard
Director, Environmental Science and
  Policy Programs and Professor
School of Public and Environmental
  Affairs
Indiana University
Bloomington, Indiana

Joy B. Zedler
Professor of Biology
Department of Biology
San Diego State University
San Diego, California

Carl E. Zipper
Senior Research Associate
Soil and Environmental Sciences
  Section
Department of Agronomy
Virginia Polytechnic Institute and State
  University
Blacksburg, Virginia

TABLE OF CONTENTS

Chapter 1
Restoration Ecology: The New Frontier 1
John Cairns, Jr.

Chapter 2
Restoration Ecology: A Synthetic Approach to Ecological Research 13
William R. Jordan, III

Chapter 3
River and Stream Restoration 23
James A. Gore and Franklin L. Bryant

Chapter 4
Abandoned Mines in Illinois and North Dakota: Toward an Understanding of Revegetation
Problems 39
Claus Grunwald, Louis R. Iverson, and Diane B. Szafoni

Chapter 5
Reclamation and Treatment of Contaminated Land...................................... 61
Michael A. Smith

Chapter 6
Techniques for the Creation of Wetland Habitat in Coal Slurry Ponds 91
Carol S. Thompson

Chapter 7
Evaluation of Strip Pits and Ponds for Physical Manipulation to Increase Wetlands and
Improve Habitat in Southwestern Indiana 115
Daniel E. Willard

Chapter 8
Salt Marsh Restoration: Lessons from California..................................... 123
Joy B. Zedler

Chapter 9
Improving Coal Surface Mine Reclamation in the Central Appalachian Region 139
W. Lee Daniels and Carl E. Zipper

Chapter 10
Restoration and Management of Ecosystems for Nature Conservation
in West Germany.. 163
Diedrich Bruns

Index 187

Chapter 1

RESTORATION ECOLOGY: THE NEW FRONTIER

John Cairns, Jr.

TABLE OF CONTENTS

I.   Introduction 2

II.   The New Frontier 2

III.   Theoretical Benefits of Restoration Ecology 3

IV.   Why Do Some Professional Ecologists Resist Involvement in the Restoration
        Process? 6

V.   Why are Environmentalists Frequently Against Developments in Restoration
        Ecology? 7

VI.   Is Restoration the Answer to All the Problems? 7

VII.   Closure of Hazardous Waste Sites 8

VIII.   Future Needs 10

References 11

I. INTRODUCTION

If a natural system is altered, its ecological role is either eliminated or substantially changed. In some cases, such as alteration due to surface mining, some formerly existing ecological characteristics may be replaced relatively quickly and others may take more than a human lifetime to restore. In the worst situation, restoration to the original condition may be impossible. In other cases, such as the construction of condominiums, high-rise buildings, football stadiums, etc., the alteration is permanent from the perspective of the human lifetime. Some artifacts may be demolished after 20 or 30 years, but a high probability exists that they will be replaced with larger more formidable artifacts rather than the original ecosystem. This article focuses on the more tractable problems, namely, restoration or rehabilitation of ecosystems subjected to a disturbance following which remedial measures may be immediately instituted. Even in cases where the intention is to build artifacts that will alter the ecosystem for tens or hundreds of years, consideration should be given to the options for alternative use when the life expectancy of the present artifact has been reached. For example, what should be done with a dam that has filled with silt? Should it be rejuvenated at great expense by dredging? If so, where will the dredged material be placed? What if the dredged material contains toxic substances? If an alternative ecosystem is more practicable, such as an aluvial plain, what might be done in locating and building the dam to make the conversion easier when the reservoir is no longer functioning as it was originally intended?

II. THE NEW FRONTIER

Restoration ecology provides an opportunity to establish a new frontier in both theoretical and applied ecology. Everyone is agog these days with the opportunities provided by the latest frontier activity in biology — biotechnology. Biotechnological developments have been made possible by some theoretical developments in the field of genetics that make alterations in the performance of species practical. These genetically altered organisms can produce materials that benefit (such as insulin), or they can reduce hazards that are threatening (e.g., toxic wastes). These are only single species; however, suppose communities of organisms are developed that are designed for specific purposes? Of course, agricultural systems represent a very limited but, nevertheless, crucial application of environmental management. However, vast areas of vegetation in power line right-of-ways are controlled by either periodic applications of chemicals or by cutting vegetation when it reaches a certain size. Cutting is extremely expensive, and chemical additions are undesirable even if applications are carefully monitored. Suppose, however, a community of organisms could be developed that would never grow above a certain height and would require no application of hazardous chemicals for control. Imagine the benefits if communities of organisms could be developed for hazardous waste sites that would both insure that the waste is immobilized (and not likely to leave the site through various routes) and simultaneously transform it to less harmful material. Alternatively, the waste might be concentrated by this special community so that it could be better utilized. Both of these examples are well within the scope of the field of restoration ecology.

Examples of restoration ecology can now be seen in the production of both marketable meat and usable energy in Africa. European cattle use only a small fraction of the vegetation in that continent, are quite vulnerable to disease, and drop fecal matter that remains compacted and pretty much unavailable since it is fairly resistant to recycling. David Hopcraft[1,2] has developed a reserve in Africa in which native organisms, such as giraffe and other native species, are being raised with far lower management costs and far more efficient use of the native vegetation than the European cattle. Not only is the number of pounds of meat per acre greatly increased and the management cost reduced, but the whole system is environmentally less damaging and is far closer to the natural system than is the case when European cattle are raised in large numbers on the same type of landscape. Although the fecal material of these native species is much more readily recycled into the natural system, it can also be collected and used to produce methane that can then be used to generate electric power or heat. This in turn may prevent trees from being used as firewood, and may diminish ecosystem damage.

There is, to my knowledge, no generally accepted definition of restoration ecology since this is a newly emerging field. However, one might say that restoration ecology is the full or partial placement of structural or functional characteristics that have been extinguished or diminished and the substitution of alternative qualities or characteristics than the ones originally present with the proviso they have more social, economic, or ecological value than existed in the disturbed or displaced state.

III. THEORETICAL BENEFITS OF RESTORATION ECOLOGY

The roots of ecology are so old that they may well predate recorded history. However, one might arbitrarily select the first edition of *Fundamentals of Ecology* in 1953 by Eugene P. Odum[3] as the emergence of the subdiscipline of ecology. This book presented ecology as an entity in its own right rather than as a subsection of another subdiscipline of biology. The book also placed ecology in perspective in relation to other fields of learning as well. As is often the case for fields in early stages of development, much of the earlier work was observational and only more recently became experimental. The late Robert MacArthur is credited by many as moving the field into the predictive stage of development. In addition, since most biologists are oriented to single species, many ecological studies still focus on that level of biological organization. Nevertheless, the use of the term *ecology* implies studies of higher levels of biological organization than single species. These studies are now beginning to emerge, as evidenced by the Hubbard Brook and Coweeta studies, as well as others.

Although the number of studies of disturbed ecosystems has increased substantially in the last 30 years (and markedly since the first earth day made such things academically respectable), the percentage of ecologists studying pristine or relatively pristine ecosystems is wildly disproportionate to the percentage of the earth's surface represented by these ecosystems. To state this more bluntly, most ecologists drive, fly, or boat past or over enormous territories in which the ecosystems have been disturbed to arrive at a small patch of undisturbed ecosystem. Without in any way denigrating the value of studies of pristine systems that furnish essential and indispensable information, it is curious that the interesting information that can be obtained from disturbed ecosystems and their restoration has only recently been given serious attention by theoretical ecologists.

Even more curious is the fact that many ecologists are carrying out research on disturbed ecosystems that have partially recovered. Sometimes they do this without being aware of the disturbance and frequently without giving events in the postdisturbance period the attention they deserve even within the limited scope of the individual research project. Three of the biological field stations where I have carried out research are on ecosystems that have suffered significant disturbance. The University of Michigan Biological Station on Douglas Lake is located on a tract that was clear-cut. This was followed by slash fires set to eliminate branches and other material then considered unmarketable. Figure 1 shows what the area once looked like, and Figure 2 shows its recent condition. The Rocky Mountain Biological Laboratory near Crested Butte, Colo. was once the scene of extensive mining and had a population over an order of magnitude greater than that resident today. The Bermuda Biological Station is located in one of the most heavily settled land masses on the face of the earth. These are not the only biological stations located on areas that have undergone major

FIGURE 1.   Damaged area around the University of Michigan Biological Station on Douglas Lake.

disturbance due to human activities. There are, in fact, many others. These few examples illustrate that many ecologists are studying stressed ecosystems; however, this might not be evident from the papers published on the investigations. Since the publications from these studies are regularly in prestigious, scholarly, peer-reviewed journals, it is abundantly clear that valuable information of interest primarily to theoretical ecologists can be obtained from studying disturbed ecosystems. However, this discussion shows that much more information can be obtained than is presently generated.

Almost every field of science requires substantive financial resources, and restoration ecology is no exception. In these financially troubled times where more and more scientists (many in university positions that depend on outside money) are chasing fewer dollars, one has no difficulty finding acknowledgment of this situation in the news media, professional journals, etc. It is surprising that the very substantial amounts of money only recently available for studies in restoration ecology have been virtually neglected by universities and colleges but not by consulting firms and research organizations primarily dependent on contracts and grants. There is little or no incentive to publish in most consulting firms; however, many individuals do so anyway. In some research organizations primarily dependent on outside funding, the incentives may be present, but they are not as great as the pressures to keep one's salary intact by obtaining still more funding before publishing the results of the last study. As a consequence, much information on restoration ecology does not appear in scholarly, peer-reviewed journals but appears in limited distribution reports in the "gray literature" — usually not subject to anonymous peer review in the same fashion as it would be in a scholarly journal. This means that the information is less readily available, and, that when it is available, one cannot place as great a trust in it as would be possible had it been subjected to a rigorous peer-review process by a dispassionate "outside" anonymous reviewer and editor. Furthermore, articles published in the peer-reviewed journals are generally criticized in the same or other journals, and mistakes are corrected publicly.

FIGURE 2.   Some restoration of area around the University of Michigan Biological Station, Douglas Lake.

No standard methods are available in the field of restoration ecology and, as a consequence, little of restoration ecology is routine. Therefore, the opportunities for basic research are enormous and funds are available from nontraditional sources that are not being exploited by the academic community as they should be. I understand not only the benefits of involvement with industry and applied problems but also the dangers, although a discussion of them is beyond the scope of this article. Also beyond the scope of this article, but important to the field of restoration ecology, is the stigma often attached by tenure and promotion committees to the source of the money obtained by a faculty member to support his or her research.

The primary advantage of restoration ecology research is that the investigator is forced to study the entire system rather than components of the system in isolation from each other. It is simply impossible to restore an ecosystem one species at a time, and one is thus forced to consider both the structural and the functional aspects of the ecosystem, including spatial relationships, species interactions, predator-prey relationships, nutrient and energy cycling, relationships between the physical, chemical, and biological components, etc. One of the most compelling reasons for the failure of theoretical ecologists to spend more time on restoration ecology is the exposure of serious weaknesses in many of the widely accepted theories and concepts of today. If the outcome of a prediction is highly uncertain, the underlying theoretical constructs are probably not sound. On the other hand, if the development and function of ecosystems are to be understood, predictive capabilities necessary for the foundation of restoration ecology must also be developed.

IV. WHY DO SOME PROFESSIONAL ECOLOGISTS RESIST INVOLVEMENT IN THE RESTORATION PROCESS?

1. Some fear that admission of even partial effectiveness in restoration will be viewed as a license for further ecological destruction in the name of progress and growth. They further fear that some of the wilderness areas, national parks, and other ecosystems that now have exceptional protection will have this protection reduced as a consequence of the feasibility of repairing damage following exploitation of various resources in these systems. Some attempts will almost certainly be made to intrude on previously protected areas on the grounds that the restoration process is now sufficiently advanced to make this possible. However, if the true cost and difficulties of restoration are made abundantly clear and responsibility for them correctly assigned, these costs will probably be a major deterrent to disturbing ecosystems of any kind. In addition, once costs become a major factor, great care will be taken during the period of disturbance to minimize the cost of restoration. As usual, the situation is one of balanced risks and benefits.

2. Many ecologists wish to be considered theoretical ecologists and distain the term "applied ecology". This is true even in land grant universities and other academic institutions where service to society is part of the mission of the institution. Even ecologists now working for consulting firms entirely on applied problems frequently hasten to tell anyone who will listen that this is merely a stop gap job until a "real position" in theoretical ecology materializes. The fears of these individuals are not without foundation. Tenure and promotion committees regularly make distinctions on the source of research funding, and a dollar from the National Science Foundation carries more weight with such committees than a dollar from a mining company. Unfortunately, traditional positions are unavailable for all of the theoretical ecologists now in the marketplace, and no substantial number of new positions in the customary institutions is likely to materialize in the foreseeable future. On the other hand, the prospect for positions in applied ecology is very heartening. It is quite possible that an ecologist doing applied research might have more opportunity to carry out theoretical work than in some of the positions with heavy teaching loads that theoretical ecologists are now accepting.

3. Ecology is a relatively new field and its predictive capabilities vary enormously. In some situations, the probability of predictions being accurate approach 100%. In other cases, predictions may be correct $< 10\%$ of the time. Precise predictions of the outcome or restoration practices will be, for some time, an area in which the outcome is highly uncertain. Naturally, ecologists are apprehensive about venturing into a field where they have less confidence in their pronouncements than they do in many other areas of ecology.

V. WHY ARE ENVIRONMENTALISTS FREQUENTLY AGAINST DEVELOPMENTS IN RESTORATION ECOLOGY?

1. The environmental movement that expanded enormously during the earth day periods has been confrontational for many years. Environmentalists have won some of their most significant victories in courts of law. Restoration ecology, if it is to be successful, requires close collaboration between the academic community, the environmental groups, regulatory agencies, and industry. The three latter groups have regularly confronted each other in court for more than a decade. Members of the academic community collaborate with each of the other three groups, often simultaneously, although individual ecologists of the academic community are mostly more or less in one "camp" (i.e., environmental group). Because the outcome of any course of action in restoration ecology is uncertain at this early stage in the development of the field, the best results will only be obtained with an attitude of mutual trust. This is difficult to ensure with the history of confrontation that has been characteristic of the group interactions for many years and is still continuing.

2. Environmentalists have been strongly protection-oriented regarding use of the environment. There are excellent historical reasons for this, as well as present ones. Restoration of damaged ecosystems focuses on multiple use, environmental management, and a variety of other practices that are alien to the protectionist point of view. However, if decision analysis is used in reaching management goals and if all major contending parties have a say about desirable environmental quality and condition, this public discussion will do more for the environment as a whole than any course of action that could be taken. At present, the environment is being protected in fragments when protecting a whole system would be better. Ideally, restoration ecology would force examination of the larger system if only because the organisms for re-colonizing damaged areas must frequently come from outside the damaged system. Also, determining what kind of restoration is possible necessitates looking at the larger system into which the restored area must fit. As a consequence, public attention must necessarily expand to geographic areas it might not otherwise consider.

VI. IS RESTORATION THE ANSWER TO ALL THE PROBLEMS?

Restoration is an admission that environmental quality was not protected! It is in many ways analogous to the relationship between preventative medicine and surgery — when damage has occurred, a method of quick repair whenever possible is reassuring, but avoiding the damage in the first place would have been better. In some cases, mismanagement, accidental spills due to faulty equipment or operator error, or deliberate damage to an environment while extracting fossil fuels or minerals will leave an ecosystem so seriously damaged that normal recuperative processes will not suffice, and management intervention is mandatory. Knowing when to intervene requires more skill and judgment than one might suppose. For example, following the wreck of the oil tanker *Torrey Canyon,* some clean-up methods appear to have caused more damage to the indigenous biota than the oil itself. In addition, an equilibrium was restored more rapidly to the natural systems where no intervention occurred than those with intervention. Careless use of suction devices, scrapers, dispersants, and the like may cause more stress to the ecosystem if improperly used than the material of the spill itself.

Damaged ecosystems can often be used for partially replacing systems lost elsewhere. In *Wetlands and Water Management on Mined Lands,*[4] abundant evidence is given that surface mined land can be used to create wetlands where none existed before. Since natural wetlands are being lost to agriculture (~80% of the loss) and for building construction and other purposes (20%) at a frightening rate, the replacement of many of their functions elsewhere has positive features. The book is quite readable, although some technical knowledge is required for some passages. The basic message of the book — highly productive artificial wetlands are used by water fowl, deer, and other organisms — is reassuring. Furthermore, artificial wetlands are extremely useful, at least in the short term, for improving water quality by acting as a trap for heavy metals and other toxicants and for adjusting the pH of the water and making other water quality improvements. There is also evidence in the same volume that overloading the systems will impair their ability to function effectively, a statement that would be platitudinous were it not for the fact that people are regularly overloading natural systems and are probably likely to do so for artificial systems. The possibility of replacing wetlands on systems where other types of ecosystems formerly existed should not diminish the efforts to protect the remaining valuable wetlands in the contiguous U.S.

VII. CLOSURE OF HAZARDOUS WASTE SITES

Although the precise number of hazardous waste sites in this country is controversial, the number is unquestionably exceedingly large. They represent a particular challenge in the restoration of damaged ecosystems — they must be restored without further hazard to human health and the environment. Several major options are available.

Option 1 — Detoxify the site and restore to original condition or some alternative ecologically stable condition. Option 1, though highly desirable, is probably possible only in areas where the hazard is minimal and the original ecological condition well known.

Option 2 — Option 1 may be technologically impossible, or the cost may be prohibitive. In some cases, collecting the waste is difficult because of leakage and other problems, and, even when it is collected, detoxification through incineration or some other means such as chemical treatment may not be either economically or technologically feasible. On these sites, the primary objectives would be to exclude the general public, migratory water fowl, and so on to reduce or eliminate the chances of inadvertent exposure to hazardous materials and to immobilize the waste so that adjacent areas are not contaminated. Avoiding contamination of groundwater is particularly important in this regard. Site restoration can be carried out so that evapotranspiration of water is enhanced, thereby reducing contamination of groundwater supplies. In addition, it may be possible to utilize some form of treatment. The rehabilitation processes in ecological terms for Option 2 are not well designed to date, but they offer a particular challenge in restoration ecology. Some of the management measures that may be taken in structuring soil profiles and the like can be adapted from those already proven effective for reclamation of surface mined areas.[5]

Option 3 — A third and economically more palatable option is to remove and treat those portions of the hazardous waste still in a relatively concentrated form (i.e., in storage containers and the soil immediately surrounding containers if there has been leakage) and to leave the partially dispersed and transformed wastes that are difficult to collect and treat on the site until natural transformations continue. Certain types of radioactive wastes with a known decay time might be particularly appropriate test cases for this strategy. In this particular situation, it will be important to determine what rehabilitation processes to use to immobilize the hazardous material on the site and to determine whether limited use is possible after the initial rehabilitation is completed. An additional requirement for this option would be to develop an ecosystem that would be compatible with the surrounding ecosystem and merge with it eventually. In Option 2, a case could be made for a totally different ecosystem than the surrounding ecosystem due to more limited access, a higher degree of hazard and risk, and the pressing need to contain or immobilize hazardous wastes. In Option 3, all these strategies might be done initially, perhaps with revegetation and associated other biota that would gradually be replaced as the hazard diminished since the transition time would be markedly less than that for Option 2.

Option 4 — The fourth option is to reduce the hazard or risk to a level considered acceptable to society and then to restore the site to either original condition or to some alternative ecosystem (such as a wetland) that would be acceptable to society. Although this may appear to be the most costly in the short term, in the long term it may prove to be the least expensive. A hazardous waste site where the chemicals are merely stored must be constantly monitored to determine that no groundwater is being contaminated or other movement is occurring in the toxic material from the site. If this is done properly for air, ground-, and surface water, and soil, the annual cost might be very great indeed. Such monitoring quite likely would be needed for a number of years. In addition, keeping humans and animals off the site is a problem. This would require physical barriers at the very least and, probably for the very hazardous sites, some type of security force. Finally, it is difficult to give absolute assurance that some natural disaster, such as an earthquake or hundred-year flood, will not render the safety measures ineffective.

Closure of hazardous waste sites represents a great opportunity and challenge for restoration ecologists. A glance at a map of the sites identified in the contiguous U.S. will show that they occur in a wide variety of climatic regions where the temperatures, rainfall, soil profiles, and biota are quite different. The types of problems posed by the hazardous wastes themselves are equally diverse. Although much remains to be done, a considerable body of information is available on the physical and chemical constraints of site closure. Ecological prerequisites and constraints are not nearly as advanced. Without in any way denigrating the complexity and difficulty of solving the physical and chemical problems, it is fair to assert that the ecological problems are at least as complex and, quite likely, more so. Given the rather high cost of chemical analyses these days, particularly for exotic chemicals in low concentrations, and the cost of boring sampling wells to plot the extent of groundwater contamination, the cost of getting suitable ecological information is almost certainly no greater than these costs and may well be less. As a consequence, it seems unfortunate that funds for the development of the ecological information base have not been available as they have for the chemical and physical information bases. Part of the responsibility for this lack of funding should be assigned to the community of ecologists who have not made as good a case for the generation of the necessary information as have the other professions.

Every site will be considered experimental for the near future in regard to the ecological component of closure of hazardous waste sites. Therefore, legislation that is too prescriptive will hamper the generation of necessary research information to improve the ecological capabilities and the rehabilitation and site closure process. Unfortunately, many industrial organizations are unwilling to spend significant amounts of money on the ecological component of hazardous site closure unless regulatory requirements demand it. In order to ensure minimal performance levels in this activity, regulations tend to be quite prescriptive. From an industrial standpoint, this position is acceptable. Industries wish to know that their competitors are incurring the same costs, and, equally important, they wish to be able to predict accurately the cost of every component of their operations. Treating the closure of each hazardous waste site as an experimental or research project almost ensures that the cost will not be identical for each site and that the precise cost of achieving the ecological objectives cannot be estimated with precision. One possible solution to this dilemma is to allow industries to accept one of three alternatives. (1) A certain set of objectives must be met (i.e., the closed site will not be hazardous to human health and the environment) and the company will be left to meet these requirements in its own way. If this can be done in a cost-effective manner that produces acceptable results at a lower cost than the other options, those companies with enough integrity to achieve this should be encouraged to do so. Of course, the criteria would have to be more explicit than those just given for illustrative purposes, but the strategy of placing the burden of reaching the objectives on the industry that created the problem is not without precedent. For example, Section 316A of Public Law 92-500 permits steam electric power plants to exceed national and state standards for thermal discharges if they can demonstrate no harm to the indigenous biota. This has proven to be ecologically successful and cost effective for industry and has generated much useful information. (2) A second option is to have highly prescriptive regulations using the best available methodology with an appreciable safety factor added for hazardous waste concentration due to the inadequacy of available information. These regulatory requirements would be updated periodically as new information becomes available, just as they have been with toxic substances. (3) The industry could agree to the site being used as an experimental site with the understanding that it would pay a sum comparable to the sum required for prescriptive Option 2 and that cost beyond that would be borne by some governmental agency, research foundation, or a consortium with a common interest in resolving this problem. This would enable society and the industry to share research costs, would enable the industry to predict the amount to be expended, would ensure equitable treatment with those industries following the prescriptive regulations, and would ensure a lower cost for the research information needed by society since only a portion, not the entire cost, would be borne by the government.

I favor Options 1 and 3 because either would generate useful research information. Even some useful information can be generated by Option 2, if performance monitoring and quality assurance information are generated during the closure process. Probably a mixture of all four options would provide the ideal information base at the lowest cost to society. Another major advantage of following this course of action is that the research will, in all cases, be carried out on already damaged sites. If industry is not involved, it may be necessary to damage additional sites to obtain the necessary research information — a course of action that should not be palatable to very many people.

VIII. FUTURE NEEDS

Effective ecosystem restoration or rehabilitation cannot be accomplished without good science. Among the many prerequisites for good science is the willingness to share existing information, to construct and test hypotheses and models, to validate predictions made from present information and make the necessary corrections when the predictions prove unsatisfactory, and to develop quality assurance programs so that errors can be corrected. Ecosystem rehabilitation and restoration, particularly when alternative ecosystems are used to replace the damaged ecosystem, require a close collaboration of industry, regulatory agencies, scientists and engineers, and the general public or its representatives. In order to accomplish the necessary tasks, a greater degree of mutual trust is necessary than now exists. Although this sounds utopian, the restoration of Lake Washington in the U.S., the Thames River and Estuary in the U.K., and numerous other examples show that this is possible. Although much of the early success in protecting the environment was the result of litigation, examples are now emerging of more funds being expended in litigation than it would cost to resolve the problem with existing methodology. Possibly equally important is the prescriptive way that laws and regulations are written. They are frequently so detailed and specific that they discourage innovative, creative approaches to solving problems. One consequence of this is that the regulated and the regulators are pitted against each other in such a way that simple solutions to obvious problems are often hampered or ignored because of legal technicalities. Courts of law are not conducive to good scientific judgments. In order to accomplish the objectives outlined in this discussion, industrial support for research projects will be essential since the traditional sources have less money than they once did and because most of the existing funds in many of the major foundations supporting biological research go to purely theoretical, as opposed to applied, problems. Some of the money now spent in litigation will have to be diverted to generating a good scientific base that should then reduce the legal battles further. A whole generation is accustomed to taking environmental problems to a court of law instead of a science court, and much effort, mutual understanding, and tolerance will be required to change direction. I am persuaded that there is no other way to address some of the problems briefly presented here.

REFERENCES

1. **Hopcraft, D.,** Productivity Comparison Between Thomson's Gazelle and Cattle, and Their Relation to the Ecosystem in Kenya, Ph.D. thesis, Cornell University, Ithaca, N.Y., 1975.
2. **Arman, P. and Hopcraft, D.,** Nutritional studies on East African heriboves. I. Digestibilities of dry matter, crude fibre, and crude protein in antelope, cattle and sheep, *Br. J. Nutr.,* 33, 255, 1975.
3. **Odum, E. P.,** *Fundamentals of Ecology,* W. B. Saunders, Philadelphia, 1953.
4. **Brooks, R. P., Samuel, D. E., and Hill, J. B., Eds.,** *Wetlands and Water Management on Mined Lands,* Pennsylvania State University Press, University Park, Pa., 1985.
5. **Parizek, R. R.,** Exploitation of hydrogeologic systems for abatement of acidic drainages and wetland protection, in *Wetlands and Water Management on Mined Lands,* Brooks, R. P., Samuel, D. E., and Hill, J. B., Eds., Pennsylvania State University Press, University Park, Pa., 1985, 19.

Chapter 2

RESTORATION ECOLOGY: A SYNTHETIC APPROACH TO ECOLOGICAL RESEARCH

William R. Jordan, III

While interest in ecological restoration has grown rapidly in recent years, especially during the decade since the passage of the Surface Mine and Reclamation Act of 1977, research in this area has generally been highly empirical in nature. In most cases, the objective has been to find techniques that are effective and economical and that can serve as the basis of prescriptions for reclamationists and restorationists working under various conditions. In other words, restoration and reclamation have generally been approached as problems that are basically *technical* in nature, the implication being that they represent a form of "applied" ecology — a task that may benefit from ecological concepts and knowledge, but that has little to contribute to it.

At the same time, it is generally recognized that the distinction between basic and applied research is not a sharp one. Information and ideas, questions and criticisms clearly do move in both directions, from the practitioner to the scientist as well as in the other direction, and work that may be carried out primarily for practical reasons often turns out to be a kind of proof-of-the-pudding test of basic ideas.

This chapter is a brief discussion of another book in which this idea is explored in some detail by some two dozen ecologists. The book was edited by myself and by Michael Gilpin of the University of California at San Diego and by John Aber of the Forestry Department at the University of Wisconsin-Madison, and will be published by Cambridge University Press early in 1987. Its title is *Restoration Ecology: a Synthetic Approach to Ecological Research*.

"Restoration ecology" is a term of our own devising and perhaps calls for a bit of explanation. By this term we do not mean simply an ecologically sophisticated or environmentally sensitive approach to the task of ecological restoration. What we mean is ecological restoration carried out specifically as a way of raising basic questions and testing fundamental hypotheses about the communities and ecosystems being restored. We mean, in other words, restoration deliberately used as a technique for basic research. The idea is simply to draw attention to the fact that one of the most powerful ways of learning about a thing, a process or a system, and one of the most convincing ways to demonstrate that you understand it, is first to take it apart, and then to put it back together and make it work. In this sense, then, restoration ecology is the second and complementary phase of experimental study: the synthesis that follows description and analysis. To make this point more clearly, and also to suggest an analogy with other disciplines such as chemistry and engineering, we have from time to time used the term "synthetic ecology".

This idea is clearly not a new one, and even in ecology it is one that probably tends to be taken for granted. At the same time, as far as we are aware, it has never been developed in detail or even very clearly articulated as a procedure, a pattern or a paradigm for ecological research. This was what we have hoped to do in *Restoration Ecology*.

The book itself amounts to a discussion, from a wide variety of points of view, of the heuristic or intellectual value of ecological restoration or synthesis. (We construe these terms broadly to include attempts to assemble communities or construct ecosystems for purely experimental purposes. Indeed, one purpose of the discussion was to identify a continuum linking various synthetic techniques that are traditional in ecology with the full-scale ecological synthesis that takes place in the field in the familiar form of "restoration".)

The starting point for this discussion was the University of Wisconsin-Madison Arboretum, where attempts to restore a collection of ecological communities native to Wisconsin and the upper Midwest have been underway since 1934. This effort, which was apparently the first attempt anywhere to restore native communities systematically and on a large scale, was initially undertaken in order to establish a collection of communities that could be used for research along traditional lines. Thus, even though it was carried out for "scientific" purposes, this project was product- rather than process-oriented, as restoration projects have almost invariably been. At the same time scientists involved with the project soon recognized that the process of restoration itself provided numerous opportunities for research. Certain difficulties encountered in the early attempts to create a tallgrass prairie *without* fire, for example, had led to classic experiments on the role of fire in prairie communities. In addition, the restoration efforts eventually led to a number of other insights into the site preferences and dynamics of plant communities. What this suggested was that, though not undertaken in quite this spirit, the restoration efforts actually represented, or could be construed as, a form of manipulative research. This was a notion of considerable practical interest to the Arboretum since it was clear that the restored communities had limited value for research along traditional lines but were ideally suited for basic research using the "restoration ecology" approach.

Fortunately, however, we were not alone in our thinking along these lines. Just about the time we were consolidating our own thinking in the course of planning for celebration of the 50th anniversary of the Arboretum in 1984, several ecologists drew attention to the heuristic value of various forms of ecological management.

In particular, Anthony Bradshaw, one of the few academic ecologists in the English-speaking world to have taken a serious interest in restoration as an intellectual as well as a practical challenge, suggested that, far from being merely a form of "applied" ecology, restoration actually represented the "acid test" of ecological understanding. About the same time John Harper, reflecting on his own early experience in agricultural research, pointed out that ecology has learned some of its most important lessons from agriculture, and that this is true precisely because the constructing and managing of ecosystems, which is the distinctive business of agriculture, has led to valuable insights and tests of ideas that are inaccessible to descriptive or even analytical studies. What Harper proposed, in effect, is that agriculture provides a paradigm for ecological research, and that the way for ecology to become experimental is by working, as farmers do, by assembling working systems, beginning with single elements to produce increasingly complex — and realistic — combinations. This, of course, was precisely what we meant by "restoration ecology".

At the same time John Cairns was drawing attention to the heuristic value of restoration efforts in a variety of situations on this side of the Atlantic. All this encouraged us to organize a symposium on "restoration ecology", as one of several events marking the 50th anniversary of the Arboretum. The symposium was held in Madison in October 1984. It included 14 formal papers and was attended by more than 300 ecologists and managers from all over the U.S. and Canada. The papers read at that meeting became the basis for the book being summarized here, though a number of additional papers have since been added. Fortunately, all three ecologists mentioned above have contributed to this effort.

Our objective throughout this project was basically to test the idea that the synthesis, repair, or healing of systems might have as much heuristic value in ecology as it does in other disciplines, such as, for example, mechanics or medicine. As we proceeded it soon became clear that there were numerous reasons for believing this to be the case. Actual restoration of communitites in the field, such as had been done at the Arboretum, is only one example of a much more general tradition of testing ecological ideas through the assembly and manipulation of communities, in the laboratory as well as in the field. In other words, what we were calling "restoration" or "synthetic" ecology was in fact a well-established approach to ecological research, even if it had not generally been recognized as such. Ecologists regularly put systems together as part of their attempts to test ideas about them, though this work rarely includes actual creation of replicas of naturally occurring communities in the field. By the same token, restorationists are continually assembling whole communities and ecosystems in the field, though rarely if ever for explicitly experimental purposes. (More precisely, experiments in restoration are typically experiments of a purely technical or empirical nature, intended to find out what works, not why it works or how the community or ecosystem itself works.)

It was our intention to see how the two traditions might be combined, or linked. The idea was to explore the idea that the tradition of synthetic research that ecologists had carried on for years, but which had usually been confined to the laboratory, might profitably be extended into the field and to work with more complex systems, modeled on those that occur naturally. In a sense we were testing an historical hypothesis having to do with the way ecological research is done. The hypothesis was that ecologists do in fact carry out various kinds of synthetic experiments, and that this suggests a pattern in ecological discovery, and also a paradigm that might be useful in planning ecological research as well as in thinking and writing about it, evaluating it, and even criticizing it.

Though this seemed a straightforward premise, we did encounter a few difficulties in our attempt to explore it. A basic problem was that it involved looking at a familiar subject from an unfamiliar direction. While recognizing the value of management work as an *occasion* for reaching new insights, both ecologists and ecological managers have tended to assume that, in the last analysis, and for most practical purposes, information, ideas, questions, and answers to questions flow in the other direction — from the theorist and the academic to the practitioner. Our goal, in contrast, was to consider restoration and synthesis not as the beneficiary of more ''fundamental'' studies, but as a technique for research, a source of questions, including questions of the most basic kind, and finally as a test — perhaps even the most stringent test — of ecological ideas, and therefore as a challenge that is not peripheral to ecology but central to it.

A second problem was that in order to do this it was necessary to approach the subject historically, which is unusual in the reporting of scientific results. A couple of generations ago, G. K. Chesterton remarked on ''the modern innovation which has substituted journalism for history'' (in: *Charles Dickens: Last of the Great Men*). But if Chesterton thought journalism was nonhistorical, it would be interesting to know what he would think of a typical modern-day scientific paper. Science, of course, has gone far beyond journalism in the elimination of history from its day-to-day communications, and has replaced the journalistic remnants of the story with something called ''materials and methods'', which, as any practicing scientist knows, is typically completely nonhistorical. (The interested reader will find further, and considerably more pungent, comments on this matter in Peter Medawar's essay ''The Art of the Soluble.'')

This tradition presented us with special problems in bringing together *Restoration Ecology,* since as the discussion and editing proceeded it gradually became clear that the whole project represented an attempt to figure out not just what ecologists know, but to get some idea of how they come to know it. Specifically the idea was to tell the story of various research efforts in order to test the historical hypothesis that a particular kind of activity — the synthetic and restorative — really has played a role in the shaping and testing of ecological ideas.

The results certainly suggest that this is the case. Detailed accounts are presented in the book. However, what may be of special interest here are several generalizations about the value of restoration as a technique for basic ecological research.

Most generally, it would seem that the greatest value of restoration as an ecological research technique is the power it has to draw attention to what is important in a system —

in other words, to force the ecologist to an increasingly clear conception of the critical parameters governing the system with which he or she is dealing. This idea is developed in a series of straightforward examples in a chapter by Anthony Bradshaw dealing with attempts to restore vegetation on profoundly disturbed land. Bradshaw first points out that plants have certain very simple needs such as the need for a suitable rooting medium, an adequate supply of water and nutrients, a lack of toxicity, and a means of reaching a site through some dispersal mechanism. He then describes in some detail how both descriptive and manipulative studies of changes in vegetation on disturbed sites provide opportunities for identifying those factors that are most significant in a given situation. Of special interest here is Bradshaw's discussion of the value of active restoration involving attempts to introduce plants and to accelerate and guide the development of a particular kind of plant community. Under these conditions, each act of disruption is likely to reveal a new set of factors crucial to the normal functioning of the system. Similarly, each manipulation of the system in an attempt to accelerate or guide its recovery is based on some kind of idea of these factors, and therefore constitutes a test of that idea.

What emerges is an elegant corroboration of Harper's idea of agriculture, with its combination of analytic and synthetic manipulations of ecological systems, as a paradigm for experimental ecology. Thus, for example, at the level of ecosystem disruption and reconstruction represented by traditional farming methods on, say, prairie soils in the American Midwest, supplies of certain nutrients such as nitrogen soon become limiting, so that the attempt to manage the resulting, artificial, agro-ecosystems initially resulted in a preoccupation with macronutrients and a considerable amount of research related to their use by the community. As Harper points out, the resulting information is of direct relevance to ecology, and — which is of special interest here — it is information that emerged as a result of efforts to take ecosystems apart, to put them back together again in a simplified, ecologically stylized form, and then to make them *work,* reliably, predictably, even economically.

It is possible to discern a pattern here: the way to understand the system is first to describe it, then to disrupt it, then to try to repair the disruption. Of course, different kinds or degrees of disruption elicit and bring into the foreground different aspects of the system. Thus, while the disruption represented by modern agriculture quickly results in failure of nutrient-supplying systems, it does not draw attention quite so forcefully or immediately to problems involving the physical properties of the soil. These questions are encountered in their most unmistakable form when the soil itself is severely disturbed, or even destroyed, as in stripping for surface mining. Thus, research on reclamation has led to numerous insights into the interrelationships between plants and soils, some aspects of which are discussed by several contributors. Bradshaw, for example, discusses ways in which reclamation work has contributed to an understanding of the importance of soil structure and its relationship to rooting depth in plant communities and Clark Ashby notes that forest restoration efforts have led to a growing awareness of the importance of macrochannels created by decayed roots and burrowing animals.

Perhaps the greatest challenge, raising the largest number of questions, is restoration of natural communities on the most profoundly disturbed sites, since it is in doing this that the restoration ecologist encounters problems related to the widest range of ecological phenomena. Mike Miller, for example, who has done extensive work involving restoration of native vegetation on minesites in the semiarid West, details how work of this kind has contributed to the understanding of mycorrhizae and the ways they interact with the vegetation in a community. He also describes a number of experiments involving the stockpiling and respreading of soil, some of which amounted to tests of various ideas about the nature and mechanism of succession in these communities. One idea, for example, was that succession beyond the early seral stages depends on mycorrhizae. This suggested that it might be possible to accelerate succession, or even to skip stages, simply by adding mycorrhizae; but when this was tried it produced negative results.

A recurring theme, running through many chapters in the book, is the value of restoration as a way of carrying the ecologist beyond what can be concluded or inferred from observation alone. Observations yield facts, but may provide little insight into the precise importance or significance of those facts. As early as the 1940s, for example, limnologists were aware of chemical reactions that might result in the release of phosphorus from lake-bottom sediments under certain conditions. Early models of lake eutrophication, based largely on experience with deep, oligotrophic lakes, represented lakes as simple ''algae bowls'', in which storage and release of critical nutrients were limited, implying that reductions in nutrient input ought to lead to corresponding decreases in productivity. According to Eugene Welch and Dennis Cooke, who contributed a chapter on lake restoration, it was not until managers started trying to restore eutrophic lakes that the weaknesses of this model were fully appreciated. Restoration efforts based on this model and involving reductions in phosphorus input were dramatically successful in certain lakes, but had little effect in others, leading directly to identification of large reserves of phosphorus in the sediments of some lakes and to a growing understanding of the factors influencing the movement of phosphorus in and out of these reservoirs.

Of particular interest here is the fact that some of the *chemistry* pertinent to the storage and release of phosphorus from lake sediments had been clearly understood since the 1940s, but that the *ecological significance* of these reactions had not been appreciated until restoration efforts drew attention to them. In fact, this is typical of the way in which restoration has served to bring to the ecologist's attention factors that may have been overlooked in the course of work with undisturbed systems.

More generally, restoration may lead to more accurate assessments of the importance of factors, the significance of which may have been under- or overestimated. The book offers a number of examples. Walter Adey's attempts to construct an artificial coral reef/lagoon microcosm in connection with his research at the Smithsonian Institution, for example, led directly to insights into the nutrient budgets and productivity of these complex systems. Similarly, by following changes in the distribution of plants on the artificial prairies at the Arboretum, Grant Cottam and colleagues have been able to show that these changes are correlated with soil moisture and pH, but not with several other soil properties, including texture and content of major mineral nutrients and organic matter. Similar observations might have been made on natural communities, but it is likely that the results would have been less revealing, since it is precisely the initial *misplacement* of plants in restoration efforts, together with the resulting tensions and unbalances and the response of the community to these, that make possible the kinds of distinctions suggested by the work Cottam describes.

In a similar way, the artificial forest communities at the Arboretum have also provided opportunities for studies of nutrient budgets that it might have been difficult or impossible to carry out in natural systems alone. Basically this reflects the difficulty of matching communities — and species — to sites. As a general rule, the restorationist attempts to make these matches as closely as possible, but inevitably slight mismatches occur, and it is the results of these, far more than the results of the most serendipitous successes, that provide insights into the ecology of the system being restored. Thus the several artificial forests at the Arboretum, which have differed considerably in performance, have made it possible for John Aber and colleagues to carry out measurements of nitrogen cycling over a far wider range of conditions than are likely to be found in naturally occurring forests. The results have led Aber to suggest a refinement in the use of the term ''mesic'' as it is used to refer to the ecological condition of a site. Generally, this term is applied to sites that are regarded as having a generous supply of both water and nutrients. By measuring productivity over the unusually wide range of conditions provided by the restored communities at the Arboretum, however, Aber and colleagues have been able to show that the sugar maple is far more sensitive to changes in mositure than to changes in nitrogen availability, making it possible to discriminate between two factors that have usually been lumped together under a single term, and in this way coming a step closer to the isolation and identification of an important parameter influencing the development of these forests.

Restoration, in other words, provides a powerful tool for discrimination in the development of ecological concepts. Another example, not discussed in the book, has to do with the definition of species. In an earlier publication ("After Description," in *The Plant Community as a Working Mechanism,* E. I. Newman, Ed., Blackwell Scientific, Oxford), John Harper pointed out that species defined in the classical manner, on the basis of morphological characteristics, may be highly inappropriate categories for ecological studies in which subtle physiological factors may be of far greater importance than even fairly well-marked morphological characters. In fact, this is a matter of some concern to restorationists who (like horticulturists) have long been aware that the behavior of individuals and populations of a species may differ across the range of the species in a way that may be highly significant ecologically. This experience clearly suggest the weakness of traditional taxonomic schemes for use in ecological work, and also suggests ways by which more ecologically oriented systems might be worked out and refined through restoration projects involving the use of reciprocal transplants and creation of communities including species and ecotypes in various, perhaps novel, combinations. Indeed, the growing use of restoration as a strategy for dealing with environmental problems lends this issue some urgency, since restoration frequently involves bringing in species from considerable distances. Here the question of provenance and the definition and ecological significance of ecotypes becomes one of considerable importance from a purely practical and economic point of view as well as from the point of view of taxonomy and evolutionary theory. A chapter by Tom McNeilly in *Restoration Ecology* includes a discussion of some aspects of this issue.

Another aspect of the heuristic value of restoration that is discussed in some detail in the book is the value of restoration as a way of working out mechanisms underlying phenomena that may have been identified in the course of descriptive studies. An example is the characterization of the mechanisms that account for changes in a community during the process of succession. Most studies of succession have been descriptive studies in which attempts have been made to discern patterns of change and, insofar as possible, to infer the reason for it by establishing correlations between specific factors and the changes taking place. The problem, of course, is that correlations can be highly misleading, and this is especially true when one is working with natural, or more or less undisturbed systems, where species and systems have long periods to form a more or less integrated whole. Under these conditions correlations may reflect quite obvious factors (e.g., water availability) or much less obvious factors, or even complex combinations of factors. John Aber's refinement of the term "mesic" suggests the value restoration has as a way of refining ideas that may seem perfectly adequate when tested by purely descriptive or even analytical studies.

In any event, it is clear that one way to test a guess about ecological systems based on observation, correlation, or analysis is to try to manipulate the system in such a way as to test the guess — and this frequently involves assembly, or partial assembly of the system. Katherine Gross' work on mechanisms of succession in old fields in southern Michigan provides an excellent example. Descriptive studies of these fields had revealed that several biennial species differed quite dramatically in the time at which they colonized these fields and also in the length of time they persist in them following colonization. These patterns of behavior were correlated with certain life-history traits such as seed size and number, the length of time the seeds remain viable in the soil, the growth form of the emerging seedlings, and so forth. In order to establish which of these factors really accounted for the behavior of the species, however, Gross undertook a series of experiments involving introduction of the plants into the community under conditions designed to test specific hypotheses accounting for their behavior. In this way she was able to show, for example, that the ability of large-seeded species such as *Daucus carota* to persist in increasingly dense vegetation longer than certain smaller-seeded species is a direct result of the large seed, which enables seedlings to become established in a vegetative cover where seedlings emerging from smaller seeds may fail.

In a similar way, to test the idea that clearings created by burrowing animals might play a key role in the persistence of certain species, Gross created artificial disturbances in the communities. In this way she was able to show that the species that colonize such disturbed areas vary depending on the season the disturbances are created.

This is obviously of considerable interest from a practical as well as from a theoretical point of view. Gross also points out, however, that the seasonal effects she observed only suggest, and may not necessarily indicate, the specific factors (i.e., the actual critical parameters, such as timing of changes in moisture supply) that are really responsible for the observed effect. This is an observation of great significance for restoration and reclamation research generally, since it draws attention to the fact that the results of purely empirical experiments are necessarily of limited applicability. What worked last spring on one site may very well not work next spring on another site, unless the restorationist actually understands the critical parameters that underly what is going on and is able to manipulate them, e.g., by irrigating at some critical moment. To identify these factors, however, it will be necessary to undertake additional synthetic experiments such as experimental irrigation of test plots, for example.

Another way of putting all this is to say that restoration provides the ecologist with a basis for distinguishing between what one really needs to know in order to understand a system and what can one afford to "black box" and, at least for the moment, ignore. Which factors, for example, determine the ability of plant species to coexist? To find out, Patricia Werner planted seeds of teasel (*Dipsacus sylvestris*) in old fields, and then attempted to relate the success of the resulting plants to the species growing near them. What she found was that the results could be accounted for almost entirely in terms of the growth form or "architecture" of the various competing species. In short, these synthetic experiments, technically almost indistinguishable from certain kinds of agricultural experiments, suggested that, so far as teasel is concerned, all that matters is the size and shape of the competition. This in turn points to the possibility of species substitutions in the creation of artificial ecological communities — a practice that may be of great importance in reconstructing communities from which a dominant species may have become extinct (e.g., American chestnut or perhaps American elm.)

A feature of this book that will be of special interest to those interested in the more philosophical aspects of ecology generally and restoration ecology in particular are several discussions in the introductory section that deal with the nature of restoration, its heuristic value and its relationship to the development of ecological theory. A key issue here is the suggestion that restoration represents the crucial, or "acid", test of ecological understanding. Plausible as this idea may seem on the surface, it is also evident that restoration or repair of a biological entity such as an ecological community may not be a stringent test of an idea simply because biological systems have a certain capacity for self-repair, so that it may be possible to "restore" a community simply by bringing elements together and letting nature take its course. Given enough time any system will recover, and this obviously proves nothing at all about anyone understanding the system.

What this means, however, is that the really critical test of understanding is not simply the ability to restore the system, but the ability to control its development and functioning. Restoration is only a particular form of control, even though it may be an especially dramatic one. By the same token, what is called for by way of an "acid test" of ecological understanding is not rote copying of natural systems (i.e., not restoration at all in the narrowest sense) but rather *imitation* of natural systems, the distinction being that copying implies reproducing the model in concrete form, while imitation implies making a similar system that only *acts* the same as the model. Clearly it is this, and not copying, that demonstrates a grasp of the critical parameters governing the system. It is in this sense, then, that we use the word "restoration" in the term "restoration ecology".

Clearly a number of questions remain. Even granting that restoration has provided ecologists with opportunities for achieving significant insights, just what has the relationship between the restoration effort and the insight been? Has it been intentional? Has it even been necessary? So far as I am aware, restoration projects have rarely, if ever, been undertaken specifically as experiments to test basic hypotheses. However, to what extent might it be desirable that this be done? One such project, which will involve planting mixtures of prairie species to test ideas about competition between them, is briefly outlined in a chapter in *Restoration Ecology,* and in fact is just getting underway outside my window here at the Arboretum as I write. One of our ideas at the outset was that laboratory experiments involving the creation of experimental communities might be extended in this way to experiments in the field; but to what extent will this prove to be practical, intellectually fruitful, or heuristically economical? Granted that questions may be answered through restoration, still there may be easier, more economical ways to answer them. Just what will happen in the process? Will phenomena such as the role of competition in shaping a community, so elegantly revealed by Michael Gilpin's experiments involving mixtures of fruit fly species in bottles, simply become irrelevant in the complex community in the field?

All this remains to be seen. At the same time, we have at least emerged from this phase of our discussion with the sense that the discussion is worthwhile. Overall, a sense emerges that the gap between theory and practice may not be so great after all. Restoration in the sense we are using it here may indeed have a great deal to offer ecology. Anthony Bradshaw has suggested that it represents the acid test of the ecologist's understanding of his or her subject. Michael Rosenzweig adds to this observation that this is so precisely because the nonlinearity of response in many ecological systems makes it necessary not only to work with perturbed systems, but to work with *profoundly* perturbed systems, in which variables such as populations sizes have been pushed to extreme values. For ethical reasons, Rosenzweig suggests this be done not by perturbing systems artificially, but by beginning with sites disturbed by activities such as mining and then adding species experimentally.

Furthermore, the idea that restoration (or "control") of the ecosystem represents the acid test of ecology is one that obviously has important implications for the science. To the extent it is valid, for example, it suggests that, far from being a mere intellectual by-product of ecology, of marginal interest to the ecological theorist, restoration actually deserves to be regarded as the organizing principle for the science, providing a basis for deciding what is worth knowing and what is not.

This immediately raises yet another question: Is this principle comprehensive enough? Or might it be possible to "restore" (i.e., to control) a system in the way we have described without really understanding it? If an ecologist can finally create a given community anywhere, at any time and under various conditions, specifying all the conditions that make this possible, is ecology then finished with that system? Or are there things ecologists properly want to know about their subject that are irrelevant to the challenge of restoration? Certainly there might be. Recently, for example, a colleague suggested that branching patterns might be an example. This is something that ecologists are interested in and that might seem irrelevant to restoration. Yet, as work like that of Pat Werner and Kay Gross clearly shows, it is not irrelevant. Still, this question awaits further discussion.

Finally, developing the idea of restoration ecology and taking advantage of it may well prove an excellent way to establish a closer, more fruitful relationship between ecological theory and ecological practice, since it suggests a common ground between theorist and practitioner. At bottom, this is the whole purpose of the discussion carried on in this book: to weaken the barrier between theory and practice by developing the idea that what the practioner has done — perhaps for purely "practical" reasons — actually represents a test of the ideas on which the procedure is based, and so, properly construed, is essentially an experiment.

Restoration, then, clearly provides ecology with a powerful technique. It may also provide it with a set of objective, definable goals, and so with a basis for self-criticism. Just as important, it provides ecology with a mission. If nothing else, restoration ecology does represent ecology taken seriously not simply as a science, but as one of the healing arts — in fact the most comprehensive of the healing arts, the one that deals not with individual members of a single species, but with the whole. It is just this sort of ecology that Aldo Leopold had in mind when, in 1940, reflecting on the state of his own discipline of game managememt, he said that it was not merely a science because it was based on value and concern for particular things. ("The state of the profession," *J. Wildlife Management,* 4(3), 343.) It is also certainly what he had in mind when he looked forward to the development of what he called a science and art of land health.

What is of special interest here is the way restoration ecology suggests that the intellectual objectives and the ethical concerns of ecology not only overlap but may actually *coinicide*. What one needs to know about the system in order to restore it may also be just the thing that is most worth knowing from a purely intellectual point of view.

To the extent this is true, it is immensely encouraging since it suggests the possibility of a science wholly integrated with value. Perhaps it is just this integration, this coincidence of interest and value that characterizes the healing arts and distinguishes them from the "mere" sciences.

In any event the discussion is an interesting one. What we have heard so far encourages us to suspect that restoration will prove a healthy preoccupation for ecology, and especially for ecologists interested in the development of a discipline that has moved beyond description and diagnosis to take on the even more demanding tasks of prediction, prognosis, and cure.

Chapter 3

RIVER AND STREAM RESTORATION

James A. Gore and Franklin L. Bryant

TABLE OF CONTENTS

I.   Introduction 24

II.   Hydrology 24

III. Water Quality 27

IV.   Riparian Vegetation 28

V.   Macroinvertebrates 30

VI.   Fish Habitat Enhancement 33

VII. Planning and Monitoring 35

References 36

I. INTRODUCTION

Streams and rivers are among the most enduring geomorphic features on the surface of the earth. These lotic environments are the primary habitats for a variety of uniquely adapted plants, invertebrates, and vertebrates. In addition to providing habitat to a diverse flora and fauna, rivers and streams have been used by humans to provide nourishment, energy, and transportation. Though rivers and streams represent a small portion of available freshwater (<0.02%), the natural replenishment of this component of the hydrologic cycle (often 2 weeks) has caused humans to perceive this resource as more available and less destructable. In turn, overuse and misuse of streams and rivers have resulted in severe damage to the ecosystems associated with running water systems.

The flow, power, and erosive capabilities of rivers have drawn people to take advantage of those hydrologic forces. Access to deposits of various mineral ores (most recently uranium and coal) has been provided by erosion of overburden materials. Until passage of the Surface Mine Reclamation and Control Act of 1977, relocation of river channels to gain access to underlying deposits was common practice. Vinikour[1] and Cairns et al.[2] documented the total loss of running water communities after alteration of flow into previously mined surface mine pits.[3]

Rivers and streams have historically provided major transportation routes along flood plains. Increased construction and demands for accessible routes have "necessitated" relocation of streams to accomodate highways, bridges, and railway roadbed. Barton[4] reported changes in benthic communities impacted by highway construction while Chapman and Knudsen[5] documented declining habitat and salmonid production in river reaches altered by flood control devices.

Ecosystem damage can also take on more subtle characteristics such as increased sediment loads, poor water quality from urban runoff, and unobserved declines of habitat and production of game and nongame species.[6-8]

With the worldwide decline in biological diversity,[9] restoration and maintenance of unique stream ecosystems is imperative. Restoration activities not only maintain a valuable ecological resource but provide the potential for a manageable, relatively renewable, resource of freshwater.

Unlike terrestrial reclamation or restoration projects, river restoration must also account for systems that interact to produce the lotic ecosystem. That is, pollutant loads from surface runoff must be reduced or eliminated and riparian vegetation must be restored to control erosion and to provide habitat for riparian fauna that also use lotic resources. Within the restored channel, a number of hydrologic considerations must be examined. Because discharge patterns are fairly predictable, channel configuration can be designed to simulate meander and depositional patterns. In addition, bank and bed stability must be planned and structures built. Both fish and their food sources, mostly from the benthic community, must also be restored. Since river restoration rarely involves direct transplantation or introduction of animal species, habitat structures should be designed to attract residents from colonizers arriving from upstream and downstream areas. Thus, "reclamation" of the instream flora and fauna is a process of habitat and colonization enhancement.

We have attempted to summarize the major structures and considerations necessary for restoration of lotic ecosystems. Some of this information is available in greater detail in agency publications[10-11] and recent texts.[12]

II. HYDROLOGY

In order for a stream restoration project to succeed, much effort needs to be made in the design and construction of a reasonably stable stream channel. Of course, stream channels are dynamic landforms, subject to change by variation in the hydraulic forces acting on the channel itself. Changes in flow can result in variations in the rates of erosion or deposition that will also affect stream channels. However, a stream geometry that can be controlled and stabilized by local geomorphic patterns is a necessity *before* implementing further restoration. The "straight ditch" channelization more often than not negates enhanced biotic recovery.

Hasfurther[13] listed four major factors that interact to establish equilibrium conditions in streams. These factors are geologic, hydrologic, hydraulic, and geometric.

Geologic factors include soil type and topography. Both of these factors influence the type and amount of sediments that enter a stream and can greatly affect the formation of meanders in a stream channel. Soil type will affect the degree of meandering through the alluvium, while certain mechanical controls may also be imposed by the relief of the area.

Stream meandering is affected by hydrologic factors as changes in stream flow and runoff. The influence of man and long-term climatic factors can greatly influence the hydrologic characteristics of a watershed. A change in vegetative cover on the watershed subsequently affects the amount of water entering a stream. A change in channel morphology can then be expected from a change in runoff characteristics.

Hydraulic factors (depth, velocity, and slope) directly control sediment transport and erosion of the stream bank. Pool and riffle formation are also controlled by hydraulic factors due to differential transport of sediments in terms of particle size and water velocities.[14]

Channel geometry involves stream pattern, channel cross-sectional shape, and the pool/riffle pattern.[13] Often these factors are affected by sediment size and discharge. For example, braided streams are generally associated with variable flow rates and large sediment particles. Bhowmik[14] reported several techniques that involved the proper design of stable stream channels by analysis of channel geometry.

Meandering is an important factor in stream channel morphology. Bloom[15] indicated that meanders are sine-generated curves that tend to minimize variance. In a meandering stream, the work done on each stream segment is uniform and total work is minimal. Bloom also stated that meanders are likely to occur due to randomness. Thus, meandering channels function very well in the transport of water and can be assumed to form naturally in fine substrates.

In order for the process of meandering to be applied in stream restoration projects, several parameters must be defined. Hasfurther[13] lists four major parameters used in quantifying meander shapes. These four parameters are wavelength, sinuosity, radius of curvature, and peak-to-peak amplitude (menader belt width). Figure 1 illustrates the meander parameters most commonly used. Hasfurther defines linear wavelength as twice the linear distance between successive inflection points, and the definition of sinuosity as the ratio of thalweg distance to arc distance (Figure 1). Radius of curvature is approximated by the radius of a circle that passes through one point and two points nearby (Figure 1). Meander belt width is the distance across the meander from successive convex peaks (Figure 1).

Several hydrologic and hydraulic parameters are also used in conjunction with the previously mentioned meander parameters. These parameters include, drainage area (A_{drain}), stream bankfull width (W), discharge (Q), sediment load index (M), stream gradient (Sch), and depth of flow (d).[13]

In order for a stream channel to remain stable, Lane[16] proposed a model in which the quantity of sediment (Qs) and the size of the sediment particles (d) is proportional to discharge of water (Qw) and the slope of the stream (S):

$$Qsd \; \alpha \; QwS \qquad (1)$$

For equilibrium to be attained following a disturbance, the values must readjust depending

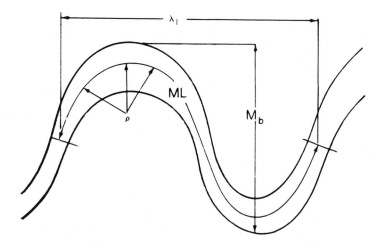

FIGURE 1.   Meander parameters commonly used in stream meander design. λ_l, linear wavelength; ML, meander length; M_b, meander belt width; ρ, radius of curvature. (Modified from Hasfurther, V. R., in *The Restoration of Rivers and Streams*, Gore, J. A., Ed., Butterworth, Boston, 1985, 21.)

upon which parameters were affected. For example, if discharge (Qw) were to decline below an impoundment then the quantity of sediment (Qs) must also decline assuming slope (S) and sediment size (d) remain unchanged. This relationship must be considered when designing a stream channel.

Meander length (M_l), radius of curvature (ρ), and sinuosity (P) can be approximated by:

$$\rho = 2.42 \text{ W} \tag{2}$$

$$M_l = 10.9 \text{ W}^{1.01} \tag{3}$$

where W is stream width (Figure 1) and

$$P \propto a \text{ M}^b$$

when M is sediment load and a ($\sim$1) and b ($\sim$0.2) are constants.[13]

The use of meanders in stream channel design appears to be a useful technique in channel stabilization. Meanders have been shown to be excellent structures in dissipating energy in stream systems.[17-18] Hasfurther[13] mentioned four methods by which a meandering pattern can be restored. These four methods include the "carbon copy" technique, an empirical approach, a natural approach, and a systems approach.

The carbon copy technique, as the name implies, involves reestablishing a stream pattern as it was prior to disturbance. When considering this approach with Equation 1, the shortcomings become obvious. To assume that the undisturbed pattern was stable and that the bed material and slope remained the same after disturbance is certainly unrealistic. A stream channel restored using this method has little chance of reaching equilibrium in a reasonable amount of time.

The empirical approach involves extrapolating the results of studies done in areas outside of the restoration area. This approach may be inappropriate in that results obtained in one area may not apply to the area of the restoration project, resulting in serious errors.

The natural approach simply involves creating a valley and allowing the stream to approach equilibrium on its own. This approach is quite unsatisfactory due to the implicit instability of the system.

The most satisfactory method is the systems approach. The systems approach involves a complete analysis of the disturbed site by meander analysis and consideration of the effects on undisturbed sites nearby. A study on the Eastern Powder River Basin in Wyoming provides a good example of the systems approach in stream channel restoration. Rechard[19] developed several equations that determined proper channel morphology of reclaimed channels in surface mined areas.

III. WATER QUALITY

The reestablishment of good water quality is a prime concern in stream restoration efforts. Ultimately the establishment of a desirable aquatic community may depend upon the attainment and maintenance of good water quality. Of course, the chemical characterisitics of undisturbed streams vary greatly from one region to another due to geological and climatic factors. Hynes[20] and Golterman[21] give good accounts of water chemistry in undisturbed rivers and streams.

The source of pollution entering a stream must be identified prior to restoration. Pollution can be from either point or nonpoint sources. Point source pollution arises from a single well-defined source such as the drainage from a surface mine. Nonpoint source pollution arises from a much larger area and is not well defined (e.g., wide scale agricultural operations). Pollutants arising from point sources are usually more concentrated; however, they are also more easily controlled due to restricted location of the source. Nonpoint source pollutants are not highly concentrated, but are not easily controlled because of the large source area and the difficulty and cost in controlling low concentration pollutants. Although nonpoint source pollutants are low in concentration, they can result in significant reductions in water quality by their large area of influence.[22]

The conditions of the drainage basin play a major role in determining the quality of water in a reclaimed stream. Streams in drainage basins covered by little or no vegetation will carry a large suspended load. Well-vegetated drainage basins result in streams carrying dissolved rather than suspended loads. In disturbed drainage basins lacking vegetation, the water chemistry will reflect the chemical characteristics of the surrounding soil. Riley,[23] investigating water quality in lakes in a surface mined region in Ohio, reported that plantings on previously exposed mine soils resulted in enough improvement of water quality to support populations of bluegill sunfish and largemouth bass in a period of 7 years from the beginning of the planting process. Prior to the plantings, the spoil area was only sparsely vegetated for a period of 31 years with very poor water quality. Riley attributed the improved water quality to an increase in the amount of organic litter and vegetative cover that reduced oxidative reactions in the spoil. Starnes and Scanlon[24] also reported increases in water quality by revegetation of a coal surface mine spoil in eastern Tennessee.

Water quality in streams flowing through disturbed watersheds or in restored streams can be protected or improved by the establishment of riparian vegetation. The width of the buffer strip of riparian vegetation necessary to protect water quality varies depending on activity in the watershed (Table 1). The recommended widths in Table 1 should be considered in stream restoration to protect water quality after reclamation of the entire disturbed area is completed.

Herricks and Osborne[22] discussed four methods of restoring water quality to a disturbed stream: isolation, removal, transfer, and dilution.

The isolation technique has limited applications in running water systems. In lakes, for example, a pollutant may be precipitated and subsequently covered by sediments and effectively isolated from the water. Due to frequent scouring of a stream bottom, isolation of a chemical pollutant is only temporary and, when uncovered, may result in pollutants of high concentration being transported downstream with severe impacts to biota.[22] However,

Table 1
RECOMMENDED WIDTHS OF RIPARIAN BUFFER STRIPS NECESSARY TO PROTECT WATER QUALITY AND AQUATIC LIFE IN STREAMS

Function of buffer strip	Recommended width
Protect water quality from logging	8 m + 0.6 m per 1% of slope
Protect water quality from logging in municipal watersheds	16 m + 1.2 m per 1% of slope
Protect aquatic life from logging	Minimum of 30 m
Protect water quality and fish	25 m plus any additional width that supports riparian vegetation
Protect streams from adverse land management practices	30 m
Protect aquatic environment	Minimum of 15 m

Modified from Brinson, M. M., Riparian Ecosystems, Their Ecology and Status, Series: FWS10BS; -81/17, National Water Research Analysis Group, U.S. Fish and Wildlife Service, Washington, D.C., 1981.

pollutants may be isolated from streams through the construction of settling basins that act as lakes, resulting in longer retention times for pollutants. Longer retention time allows pollutants to settle before water enters a stream, avoiding the problems associated with scouring and facilitating removal efforts.

The removal technique involves either removal of a pollutant at the source or removal from the stream itself after introduction. Certainly removal at the source is preferred and offers the advantage of high concentration removal techniques that are more efficient. Removal of a substance from a stream is difficult because low concentration removal techniques are inefficient and costly. Also, removal of a substance from a stream may result in serious mechanical impact by the removal activities.

The transfer of a pollutant involves simple removal of a substance downstream by stream flow. Although this method results in the elimination of a pollutant from an individual stream reach, it presents problems to reaches downstream.

Dilution of substances in flowing water is another method of restoring water quality to disturbed streams. Substances are diluted downstream by increased water volume. Also, substances may become subject to elimination from the water as the physical characteristics of the stream change downstream;[22] e.g., an acid stream may be buffered as water flows over an exposed limestone bed.

A holistic approach in restoring a disturbed area is necessary to insure high water quality in streams. The proper approach must include the establishment of adequate vegetative cover on the disturbed watershed and particularly an adequate buffer strip of riparian vegetation.

IV. RIPARIAN VEGETATION

A stable zone of riparian vegetation enhances the water quality of a stream by reducing erosion and the subsequent introduction of suspended solids. In low-order streams, riparian vegetation provides shading that reduces water temperature and primary production and also provides organic material in the form of detritus, which is important to stream community dynamics.[25] Riparian vegetation also represents important habitat for terrestrial organisms, particularly in arid regions.[26] Riparian vegetation functions in bank stabilization, which is of particular importance in stream reclamation efforts.

For a successful riparian community to become established, an adequate substrate must be provided. Physically, the soil must be of a texture to allow root penetration by plants and have a sufficient water capacity to support vegetation. The chemical properties of the soil must also be considered when establishing riparian vegetation. Often, particularly in the case of mine spoils, the chemical composition of the soil is such that very few plants can survive.

The characteristics of reconstructed soil are often determined by the reconstruction process rather than natural soil-forming processes. The process of reconstructing soil generally involves backfilling of the stored overburden, contouring, replacement of topsoil, fertilizing, and disking. Excessive grading of the subsoil by heavy equipment can result in a virtually solid layer of material, a condition termed "massive", at the upper boundary of the subsoil.[27] McSweeney and Jansen[27] reported that root penetration by corn through a massive subsoil was restricted to desiccation cracks at the boundary between topsoil and subsoil. McSweeney and Jansen used the term "fritted" to describe a subsoil with more desirable characteristics for plant root formation. Fritted subsoils occur as a result of unconsolidated material being broken into pieces during the handling process. When in place, fritted subsoils have many gaps that facilitate root penetration. Gavande et al.[28] reported that the mixed overburden alone appeared to be sufficient to allow growth of native woodlands in reclaimed surface lignite mines in the Gulf Coast region of North America. However, due to the unconsolidated nature of the overburden, water and wind erosion were prevalent. The amount of tillage, which involves mixing the soil, performed on the reclamation area is associated with the growth rate of transplanted trees. Anderson and Ohmart[29] reported higher rates of growth for cottonwood and willow trees in tilled rather than untilled soils.

The chemical characteristics of reclaimed soil is highly dependent upon the characteristics of the overburden. In reclaimed soils associated with surface coal mining, several chemical characteristics may be observed. Barnhisel[30] reported low levels of phosphorus, nitrogen, and possibly potassium. Postmined soils are also often very acidic due to the oxidation of sulfide minerals.[24,30] High acidity and low levels of important plant nutrients in the soil can be corrected by liming and fertilizing. However, in order to determine proper soil treatments, other chemical characteristics of the reclaimed soil must be examined, preferably in the planning stages of the reclamation project.

Upon completion of soil reconstruction and fertilization, revegetation should begin. Considerations in selecting which plant species to use should include rate of growth, drought resistance, tolerance to the chemical characteristics of the reclaimed soil, value to wildlife, and aesthetic appeal. Chironis[31] provided examples of plants that meet some of the above requirements. Brinson[32] listed many species of plants commonly found in riparian communities within different regions of the U.S. which may be considered in a planting scheme.

Biological considerations must also be taken into account when revegetating reclaimed areas. Competition between plant species for water and nutrients must be reduced to allow the establishment of favorable plant species. Anderson and Ohmart[33] reported that mortality of transplanted saplings ranged from 5 to 71% when grown in association with invading weeds. However, when the weeds were removed, they reported sapling mortality from 0 to 7%.

In arid regions, irrigation may be necessary to reduce the mortality rate of transplanted trees and shrubs. Anderson and Ohmart[33] reported that irrigation for as few as 90 days was satisfactory to insure the survival of transplanted trees. Irrigation methods certainly vary; however, Anderson and Ohmart give descriptions of two different drip irrigation systems that gave adequate results.

Where water is a limiting factor, planting should begin in winter.[33] Winter planting allows the trees to become well established before water availability becomes critical. In more humid climates, planting could be performed throughout the year.

Transplanting woody species appears to be an excellent method of reestablishing woody vegetation on disturbed sites.[34] However, transplanted trees should be obtained from areas near the reclamation site. Anderson and Ohmart[33] reported good results using cuttings from wild stock started in a nursery.

Riparian areas represent very important communities and should be protected if possible. However, when economic pressures necessitate the destruction of these valuable areas, they must be reclaimed. The value of riparian vegetation to wildlife, both terrestrial and aquatic, is evident.

V. MACROINVERTEBRATES

Benthic macroinvertebrates represent a critical pathway for the transport and utilization of energy within lotic communities. The distribution and zonation of invertebrates is a continuum from headwaters to mouth as a function of transformation and availability of allochthonous input (particulates) and autochthonous production[25] and the hydrologic state of the stream and instream changes in channel hydraulics.[35] Alterations of the physical features of rivers and streams have the potential to alter energy dynamics in downstream areas and reduce available habitat.

The lotic invertebrates, especially aquatic insects, are an important component of the diets of fish species. Among salmonids, Allan[36] estimated as much as 40% of brook trout (*Salvelinus fontinalis*) diets were benthic invertebrates, while Elliot[37] found even higher proportions. Most of these food items were taken from drifting individuals in the water column. Benthic feeders (forage fish such as sculpins, *Cottus* sp.; and darters, *Etheostoma* sp. and *Percina* sp.) may utilize macroinvertebrates for their entire diets.[38] Drift distance for most aquatic invertebrates average 10 m and vary in frequency during certain life cycle stages.[39,40] Habitat restoration must be sufficient to attract invertebrate colonizers from an unstressed source area to establish a stable benthic community.

In general, there are four primary sources[41] of colonizers:

1. Drift of organisms from upstream source communities
2. Upstream migration of benthic invertebrates
3. Movement from within the substrate or adjacent bank storage areas
4. Colonization from aerial sources (oviposition by adult insect species)

Drift is a relatively continuous process in stream ecosystems. Townsend amd Hildrew[42] found that 82% of macroinvertebrate movement was a result of drift within the water column of the stream. Drift is primarily a diurnal event in any stream ecosystem containing a viable invertebrate community. A variety of factors can contribute to aquatic insects entering drift including dorsal light cues, internal physiological changes, changes in local hydraulic habitat to less than preferred conditions, depletion of local food resources, and increases in predator pressure.[43-47] Regardless of the factors that determine drift, an undisturbed upstream source area of colonizers must be maintained for most efficient restoration of impacted invertebrate communities.

Upstream migration is also important to the establishment of new communities. Although rates of arrival vary temporally, as much as 20% of new colonizing invertebrates may arrive from upstream movement.[41] Gore[45] reported upstream movements of odonate nymphs at distances of 40 km in 6 weeks or less. Thus, downstream sources of colonizers must also be maintained.

In areas of unconsolidated, porous substrates and flood plains, within-substrate migration can contribute to the establishment of new communities. This is most common in impacted streams that are temporarily dewatered but retain bank storage and/or interstitial water. In cases of diversion activity, the need for a buffer strip (for water storage and substrate stability) is indicated for maintenance of recolonizers.

In general, egg deposition activity by flying adult insects is thought to compensate for the downstream drift of nymphal and larval forms.[20] Aerial contribution is limited to the times of greatest adult emergence.[41] The upstream flight of adults suggests, again, the need for assuring a stable downstream community as a source of colonizers.

Since drift and aerial colonization are temporally varied, usually highest in spring and autumn, restoration efforts should be directed so that enhanced habitats are available during these time periods.

A critical element in restoring macroinvertebrate habitat is the ability to reproduce key habitat characteristics of the source area of colonizers. Although water quality and sources of food energy must be considered, within-site distributions are generally determined by substrate, velocity, and depth characteristics.[35,48]

Substrate composition is the most easily manipulated habitat characteristic in restoration projects. It will be necessary to consider such factors as degree of embeddedness of particles, size of particles, contour of the substrate, and heterogeneity of substrate types in the source and recipient areas.[49] Aquatic invertebrates are associated with a wide variety of substrates ranging from rooted vegetation and dead wood, to periphyton to all sizes of inorganic particulates (silt, sand, gravel, and cobble). Merritt and Cummins[50] have listed major habitat types for families and genera of aquatic insects in North America. This can be a useful guide if source areas of colonizers are unknown or unrestored. In most cases, highest diversities and production have been reported from channels with medium cobble (256 mm diameter) and gravel substrates. Of course, this is a relative estimate and can be modified according to local conditions of the source area of colonizers.

In an unregulated stream channel, it is not possible to control discharge patterns. However, hydrologic restoration must account for the normal hydrograph in maintenance of meander patterns and pool/riffle frequency. Instream structures can be used to modify local hydraulic conditions to present preferred habitat to benthic invertebrates. In most cases, these structures serve a dual purpose for fish habitat as well. Log-drop structures improved fish habitat by trapping sediment with minor impact on invertebrate densities while gabion deflectors increased water velocity over riffles to remove accumulating sediments from cobble substrates.[51] For macroinvertebrates, then, primary considerations include formation of sediment-free riffle areas and structures to control sediment deposition. Such structures as deflectors (gabion, single, or double wing), check dams, and placement of large boulders have all been demonstrated to produce scoured areas of riffle habitat and pools.[3,52,53] Gore and Johnson[3] demonstrated a transect-hydraulic method for computer mapping changes in available habitat with changes in discharge before and after placement of structures. In these cases, gravel bars created by scour and substrate particle displacement as well as man-made riffles created optimum habitat conditions. Gore and Johnson[3] and Thompson[53] pointed to the need for a pool control area within the reclaimed reach to dissipate increased sediment loads and assure maintenance of downstream communities.

The final goal of restoration is to determine if stable communities have been established in the restructured channel. The substrate particles and structured areas act as "islands" to be colonized by benthic invertebrates. The period of time to attain equilibrium can be considered a function of the distance between the source of colonizers and the "recipient" island.[54] Colonization patterns of rapid invasion, a peak diversity and density, and decline to an equilibrium state (resulting from competition and extinction) have been reported many times for aquatic insects. Maximum densities have been reported to occur between 70 and 150 days, while periods of up to 500 days have been determined for establishment of stable benthic communities.[55,56] Stability is most easily determined as a comparison of reclaimed areas to source areas by means of a taxonomic similarity index.

Prediction of rates and trends to establishment of stable communities depend on a number of factors.

1.   Gore[55] demonstrated the effect of distance on colonization times. For increases of

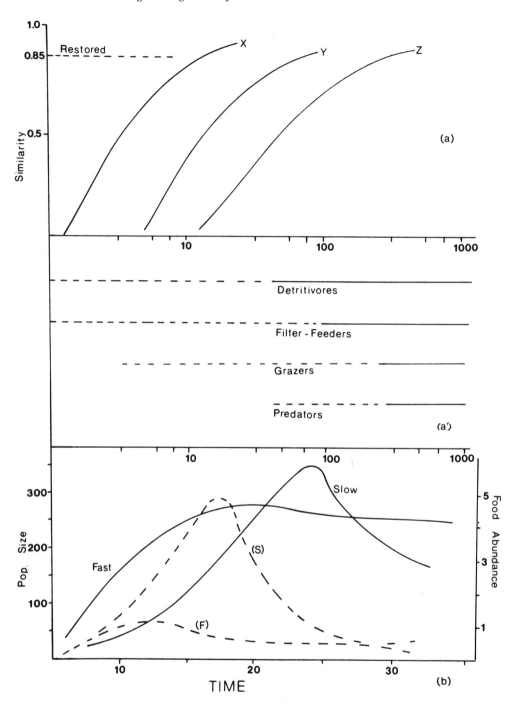

FIGURE 2.   Colonization patterns of aquatic invertebrates in restored river channels as measured by comparisons of community similarity between successive downstream stations (X, Y, Z) and source areas (a), order of arrival of main feeding groups (solid line indicates levels equal to source area) (a'), and resource tracking of slow and fast colonizers (solid line) according to food availability (dashed line) (b).[55,57] Note: time scales for (a) and (a') are logarithmic.

200 m from the upstream source area, there was a lag of 75 days to attainment of maximum densities and diversities (Figure 2a). This doubling effect was maintained for communities farther downstream and implies a discrete length of stream section

that can be restored at a single time (6.5 years for 1.5 km of stream) with no ameliorating tributary waters. Sheldon[57] modeled this "wave" of colonization effect to include temporal variations in food availability and the dispersal abilities of colonizers (high for detritivores, low for predators; as reported by Gore[55]) (Figure 2b).

2. Control of sediment deposition or siltation will increase colonization rates and enhance biotic interactions.[48]
3. Pratt and Cairns[58] have shown that the maturity (diversity and stability) of the source affects the rate of colonization. Maintenance of stable source areas resulted in attainment of communities most similar to the source areas in the shortest time.
4. Varying discharge rates must be accounted for in planning restoration. Although low discharge results in increased drift rates, colonization is not enhanced.
5. Drift rates vary temporally, as mentioned previously. In combination with controlled or predictable discharge patterns, a restoration specialist should be able to predict the most effective time for reclamation and habitat enhancement. Bogatov[59] has suggested techniques for predicting drift distances under varying flows to aid in these predictions.

The benefits of maintaining benthic communities result in stable ecosystem dynamics and provide a link to the fish assemblage of the stream ecosystem. Benthic community restoration requires minimal capital investment and the least sophisticated structure development for a return that benefits a complete restoration package.

VI. FISH HABITAT ENHANCEMENT

The most important components for fish production are acceptable water quality, food production areas, spawning/egg incubation areas, and cover.[52] Water quality needs and habitat enhancement for food production (macroinvertebrate communities) have been described in earlier sections.

Spawning habitat has been measured extensively with special emphasis on salmonid fisheries. Generally, riffle areas with velocities between 15 and 90 cm/sec, water depths >15 cm, and cobbled and graveled substrates (0.6 to 7.6 cm diameter) are preferred by most cold-water fish.[52]

A variety of chemical, hydraulic, and physical conditions must be maintained for successful incubation of eggs. Demand for oxygen increases with embryonic development but must be at least 5.0 mg/ℓ, measured interstitially. Since water percolation brings oxygen for incubation, the shape of the stream bed is critical to supplying optimum oxygenation. Vaux[60] found increased permeability and a convex stream bed induced downwelling of water into egg nests within or on the substrate while concave stream beds caused a general upwelling. The downwelling provides oxygen as well as removal of metabolic wastes. Permeability can be considered high when substrate is <5% sand and silt (as passed through a 0.833 mm sieve) and low when >15% fines.[61] Buffer strips and other activities to reduce erosion are necessary to control silting of egg nests.

Cover is any structure that protects fish from mechanical damage by high current velocities and predation. Natural cover is provided by overhanging riparian bodies, submerged macrophytes, and submerged objects. Overhead cover is utilized by species showing photonegative behavior and thigmotaxis. Wesche[62] developed a technique for classifying cover at varying discharges for intra- and inter-site comparisons. Generally, fish establish territories around cover structures that are used for resting and feeding.

A review of restoration literature has indicated that the most common structures for fish habitat enhancement have been current deflectors, overpour structures (dams and wiers), bank covers, and boulder placements. Wesche[52] has described construction details and installation characteristics. These are summarized in Table 2.

Table 2
INSTREAM HABITAT STRUCTURES TO ENHANCE ESTABLISHMENT OF FISH ASSEMBLAGES

Structure	Utility	Siting criteria	Ref.
Deflectors	Redirecting current	Streams of various size	63, 64, 70
	Stabilizing thalweg	Gradients <3%	
	Scouring pools	Bank stability opposite deflector	
	Silt removal	Alternating banks, 5—7 channel	
	Erosion abatement	widths for senicous flow	
	Consolidating low flows	Anchorer into bank 1.5 m	
	Increased pool/riffle ratios	Efficient with natural materials	
Dams	Pool formation/control	Low end of steep break in gradient	64, 67, 71
	Holding spawning gravels in up-	Stable substrate and banks	
	stream areas	Anchor into bank 2 m	
	Fish passage	Successive structures, 5—7 channel	
	Sediment control	widths	
	Collection of organic debris	Heights < 0.3 m	
		Spawning gravels between structures	
		if no passage	
Boulder placement	Added rearing habitat	During low flow	70, 72
		0.6—1.5 m diameter	
	Cover	Granitic types preferred	
	Restore meanders	Embedded a short distance	
		Greatest effect in reaches with	
		<20% pool area	
		Natural materials are most	
		economical	
Trash catchers	See dams (above)	Small, headwater streams	73
		High gradient	
		1/3 cost of log dams	

Modified from Wesche, T. A., in *The Restoration of Rivers and Streams,* Gore, J. A., Ed., Butterworth, Boston, 1985, chap. 5.

Wesche reported numerous cases of increased production and recruitment within a year of placement of deflectors. Current deflectors are built of various combinations of logs, rocks, gabions, and wire mesh. Rock-boulder deflectors have a long life span in high gradient streams. Although of little aesthetic value, gabions have the advantage of withstanding varying discharge and providing some undercut cover.[63] Double-wing deflectors are recommended in areas where thalweg stability, erosion control, and high gradient structures are required.[64] Streams can be narrowed to 80% by this technique. Underpass deflectors (a log emplaced perpendicular to flow, a few centimeters above the substrate) can be used for scour pools and sediment control in low-order streams.

Dams (check dams, wiers, and plunge dams) are most often employed to enhance habitat in high gradient, headwater streams. Duff,[65] in a summary of restoration techniques, concluded that dams have the greatest potential as fish habitat enhancement structures. Pools created by dams increase available cover, resting, and feeding opportunities for most fish species. Maughan et al.[66] found that wooden dams continued to be serviceable even after 30 years of emplacement. Low-profile dams are commonly of rock-boulder, log, or gabion construction. Rock-boulder dams are the most easily constructed. Dams up to 5 m width can be constructed in 3 to 4 hr.[67] Single log dams (low-order streams), the wedge dam, K-dam, and plank and board dams (latter two in medium-order streams) seem to have the greatest potential.[65] Rip-rap must be provided on upstream and downstream sides to prevent undercutting. Minimum log size should be 30 cm in diameter to allow at least a 15-cm water drop for scouring. In headwater streams, introduction of beaver populations to construct natural low profile dams had the effect of increasing fish production to levels greater than by construction of artificial structures.[68]

Individual and clustered boulders are the simplest in-channel treatments for fish habitat enhancement. Boulders placed in clusters should be of an upstream "V" or downstream "V" and placed away from unstable bank areas to avoid increased erosion and sediment loads to the managed areas. In streams inaccessible to heavy equipment, Cooper and Wesche[63] tested artificial boulders (constructed from gabion materials) that withstood discharges of 28 m³/sec.

Log and board overhangs, metal or fiberglass ledges, tree retards ("snags"), and rip-rap have all been used to enhance fish production in restored river and stream channels. Most often cover structures are placed for use with current deflectors and low-profile dams in low-order streams and with current deflectors and boulder placements in higher order streams. Cooper and Wesche[63] reported that sheet metal overhangs painted flat black or mottled brown were aesthetically pleasing and cost efficient since their reinforcing bar anchors allowed the shelves to be employed in a variety of bank conditions. Tree retards have the added advantage of providing bank stability in addition to cover. Trees of at least 15 cm diameter and 9 to 18 m length with trunks anchored by cable to a point 2 m from bank edge have been successfully employed in newly channelized reaches. When the crowns of the trees are quite bushy (e.g., conifers) and are combined with rock rip-rap, stable, well-covered pool areas can be maintained.[3,52] Rip-rap treatment is most successful using angular rocks that are nonerodable (0.1 to 0.8 m diameter). Slope and depth of rip-rap embankments should be considered under the advisement of available civil engineers.

VII. PLANNING AND MONITORING

The restoration of an altered stream or river ecosystem requires an interdisciplinary approach. This requires that the abilities of aquatic and riparian biologists, soil scientists, hydrologists, and civil engineers must be utilized in concert. The first steps, conducted by biologists, are the identification of species distribution, population sizes, and community structures of biota in upstream and downstream natural areas (sources of colonization). From this information, preferred habitat characteristics can be predicted. The hydrologists and engineers must evaluate the new channel to assess required shape, dimensions, and meander patterns to accommodate the normal hydrograph. Together, biologists, hydrologists, and engineers will then evaluate the hydraulic effects of habitat enhancement structures on pool/riffle frequency, bed mobility, and suspended load. Riparian specialists must also interact with engineers to assure that rip-rap structures and bank slope will be suitable for revegetation. The potential impacts of revegetation activity must also be considered. Finally, the aspects of personnel requirement, availability of natural materials for structures, equipment needs, and cost-benefit problems must be discussed with agencies supporting the project.

Monitoring of reclamation efforts is a critical phase that is often ignored. Emplacement of habitat enhancement structures and proper hydrologic structures do not always equal successful restoration. Continued monitoring of biotic communities will allow the restoration team the opportunities to further ameliorate impacts or to alter structures to further enhance biotic recovery. Winget[69] has suggested a rapid assessment technique that includes evaluation of some hydrologic characters, water quality, and invertebrate community structure to indicate success or the need to further enhance recovery in the stream ecosystem. Gore[55] suggested that simultaneous comparison of communities in the restored area with those of adjacent source areas of colonizers (presumably in natural condition) by use of a simple similarity index (Jaccard's) can indicate achievement of stable systems. This sort of comparison is often more accurate than the assumption of success based upon achievement of maximum densities or diversities. Through successful monitoring, the restoration team can rectify past mistakes and progress to more effective reclamation of future impacted river and stream ecosystems.

REFERENCES

1. **Vinikour, W. S.,** Biological consequences of stream routing through a final-cut strip-mine pit: benthic macroinvertebrates, *Hydrobiologia,* 75, 33, 1980.
2. **Cairns, J., Jr., Dickson, K. L., and Herricks, E. E.,** *Recovery and Restoration of Damaged Ecosystems,* University Press of Virginia, Charlottesville, 1977.
3. **Gore, J. A., and Johnson, L. S.,** Strip-mined river restoration, *Water Spectrum,* 13, 31, 1981.
4. **Barton, B. A.,** Short-term effects of highway construction on the limnology of a small stream in southern Ontario, *Freshwater Biol.,* 7, 99, 1977.
5. **Chapman, D. W. and Knudsen, E.,** Channelization and livestock impacts on salmonid habitat and biomass in western Washington, *Trans. Am. Fish. Soc.,* 109, 357, 1980.
6. **Newbold, J. D., Erman , D. C., and Roby, K. B.,** Effects of logging on macroinvertebrates in streams with and without buffer strips, *Can. J. Fish. Aquat. Sci.,* 37, 1076, 1980.
7. **McElroy, A. D., Chiu, S. Y., Nebgen, J. W., Aleti, A., and Vandergrift, A. E.,** Water pollution from nonpoint sources, *Water Res.,* 9, 675, 1975.
8. **Burgess, S. A. and Bides, J. R.,** Effects on stream habitat improvements on invertebrates, trout populations, and mink activity, *J. Wildl. Manage.,* 44, 871, 1980.
9. **Wilson, E. O.,** The biological diversity crisis, *BioScience,* 35, 700, 1985.
10. **Nelson, R. W., Horak, G. C., and Olson, J. E., Eds.,** Western Reservoir and Stream Habitat Improvements Handbook, FWS/OBS-78-56, U.S. Fish and Wildlife Service, Washington, D.C., 1978.
11. **Stream Enhancement Research Committee,** Stream Enhancement Guide, Province of British Colombia, Government of Canada, Ministry of Environment, Vancouver, 1980.
12. **Gore, J. A., Ed.,** *The Restoration of Rivers and Streams,* Butterworth, Boston, 1985.
13. **Hasfurther, V. R.,** The use of meander parameters in restoring hydrologic balance to reclaimed stream beds, in *The Restoration of Rivers and Streams,* Gore, J. A., Ed., Butterworth, Boston, 1985, 21.
14. **Bhowmik, N. G.,** Hydraulic considerations in the alteration of streams, in *Fish and Wildlife Relationships to Mining,* Starnes, L. B., Ed., Water Quality Section, American Fisheries Society, Milwaukee, Wis., 1985, 74.
15. **Bloom, A. L.,** *Geomorphology: A Systematic Analysis of Late Cenozoic Landforms,* Prentice-Hall, Englewood Cliffs, N.J., 1978.
16. **Lane, E. W.,** The importance of fluvial morphology in hydraulic engineering, *Proc. ASCE,* 21, 745, 1955.
17. **Leopold, L. B. and Wolman, M. G.,** *River Channel Patterns: Braided Meandering and Straight,* Geol. Surv. Prof. Paper, 282-B, USGS Neston, Va., 1957, 39.
18. **Schumm, S. A.,** *The Fluvial System,* John Wiley & Sons, New York, 1977.
19. **Rechard, R.,** Suggested stream pattern restoration for the Eastern Powder River Basin, in *Proc. 2nd Wyoming Mining Hydrology Symp.,* University of Wyoming, Laramie, 1980.
20. **Hynes, H. B. N.,** *The Ecology of Running Waters,* Liverpool University Press, England, 1970.
21. **Golterman, H. L.,** Chemistry, in *River Ecology,* Whitton, B. A., Ed., University of California Press, Berkeley, 1975, 39.
22. **Herricks, E. E. and Osborne, L. L.,** Water quality restoration and protection in streams and rivers, in *The Restoration of Rivers and Streams,* Gore, J. A., Ed., Butterworth, Boston, 1985, 1.
23. **Riley, C. V.,** Ecosystem development on coal surface-mined lands, 1918—75, in *Recovery and Restoration of Damaged Ecosystems,* Cairns, J., Jr., Dickson, K. L., and Herricks, E. E., Eds., University Press of Virginia, Charlottesville, 1977.
24. **Starnes, L. B. and Scanlon, D. H.,** Results of remedial reclamation treatments on terrestrial and aquatic ecosystems in Ollis Creek watershed, Tennessee, in *Fish and Wildlife Relationships to Mining,* Starnes, L. B., Ed., Water Quality Section, American Fisheries Society, Milwaukee, Wis., 1985, 14.
25. **Vannote, R. L., Minshall, G. W., Cummins, K. W., Sedell, J. R., and Cushing, C. E.,** The river continuum concept, *Can. J. Fish. Aquat. Sci.,* 37, 130, 1980.
26. **Hubbard, J. P.,** Importance of riparian ecosystems, in Importance, Preservation, and Management of Riparian Habitat: A Symposium, Fort Collins, Colo., USDA Forest Service, Gen. Tech. Dept. NM-43, 1977, 14.

27. **McSweeney, K. and Jansen, J.,** Soil structure and associated rooting behavior in minesoils, *Soil Sci. Soc. Am. J.,* 48, 607, 1984.
28. **Gavande, S. A., Holland, W. F., Grimshaw, T. W., and Wilson, M. L.,** Overburden management and vegetation in the Gulf Coast bignite region: problems and solutions, in *Proc. Symp. Surface Mining Hydrology, Sedimentology, and Reclamation,* University of Kentucky, Lexington, 1979, 293.
29. **Anderson, B. W. and Ohmart, R. D.,** Revegetation for Wildlife Enhancement Along the Lower Colorado River, USDI Bureau of Reclamation, Boulder City, Nev., 1982.
30. **Barnhisel, R. I.,** Reclamation of Surface Mined Coal Spoils, Dept. EPA-60017-77-093, U.S. Environmental Protection Agency, Interagency Energy Environment Research and Development Program, Washington, D.C., 1977.
31. **Chironis, N. P.,** Guide to plants for mine spoils, in *Coal Age Operating Handbook of Coal Surface Mining and Reclamation,* Chironis, N. P., Ed., Coal Age Mining Informational Services, McGraw-Hill, New York, 1978, 282.
32. **Brinson, M. M.,** Riparian Ecosystems, Their Ecology and Status, Series: FWS10BS; -81/17, National Water Research Analysis Group, U.S. Fish and Wildlife Service, Washington, D.C., 1981.
33. **Anderson, B. W. and Ohmart, R.D.,** Riparian revegetation as a mitigating process in stream and river restoration, in *The Restoration of Rivers and Streams,* Gore, J. A., Ed.., Butterworth, Boston, 1985, 41.
34. **Frizzell, E. M., Smith, J. L., and Crofts, K. A.,** Transplanting native vegetation, in Surface Coal Mining Reclamation Equipment and Techniques, Information Circular 8823, U.S. Bureau of Mines, Washington, D.C., 1980, 48.
35. **Statzner, B. and Higler, B.,** Questions and comments on the river continuum concept, *Can. J. Fish. Aquat. Sci.,* 42, 1038, 1985.
36. **Allan, J. D.,** Determinants of diet of brook trout (*Salvelinus fontinalis*) in a mountain stream, *Can. J. Fish. Aquat. Sci.,* 38, 184, 1981.
37. **Elliot, J. M.,** The food of brown and rainbow trout (*Salmo trutta* and *Salmo gairdneri*) in relation to the abundance of drifting invertebrates in a mountain stream, *Oecologia,* 12, 329, 1973.
38. **Page, L. M.,** Handbook of Darters, TFH, Neptune City, N.J., 1983.
39. **McLay, C.,** A theory concerning the distance traveled by animals entering the drift of a stream, *J. Fish. Res. Board Can.,* 27, 359, 1970.
40. **Waters, T. F.,** The drift of stream insects, *Am. Rev. Entomol.,* 17, 253, 1972.
41. **Williams, D. D. and Hynes, H. B. N.,** The recolonization methods of stream benthos, *Oikos,* 29, 265, 1976.
42. **Townsend, C. R. and Hildrew, A. G.,** Field experiment on the drifting, colonization, and continuous redistribution of stream benthos, *J. Anim. Ecol.,* 45, 759, 1976.
43. **Hughes, D. A.,** On the dorsal light response in a mayfly nymph, *Anim. Behav.,* 14, 13, 1966.
44. **Ciborowski, J. J. H.,** The effects of extended photoperiods on the drift of the mayfly *Ephemerella subvaria* McDunnough (Ephemeroptera: Ephemerellidae), *Hydrobiologia,* 62, 209, 1979.
45. **Gore, J. A.,** Reservoir manipulations and benthic macroinvertebrates in a prairie river, *Hydrobiologia,* 55, 113, 1977.
46. **Bohle, V. H. W.,** Beziehungen zwischen dem nahrunsangebot, der drift und der raumlichen verteilung bei larven von *Baetis rhodani* (Pictet) (Ephemeroptera: Baetidae), *Arch. Hydrobiol.,* 84, 500, 1978.
47. **Walton, O. E., Jr.,** Invertebrate drift from predator-prey associations, *Ecology,* 61, 1486, 1980.
48. **Gore, J. A.,** Development and application of macroinvertebrate instream flow models for regulated flow management, in *Regulated Streams: Advances in Ecology,* Kemper, B. and Craig, J. F., Eds., Plenum Press, New York, 1987, 99.
49. **Gore, J. A.,** Mechanisms of colonization and habitat enhancement for benthic macroinvertebrates in restored river channels, in, *The Restoration of Rivers and Streams,* Gore, J. A., Ed., Butterworth, Boston, 1985, chap. 4.
50. **Merritt, R. W. and Cummins, K. W., Eds.,** An Introduction to the Aquatic Insects of North America, 2nd ed., Kendall/Hunt, Dubuque, Iowa, 1984.
51. **Luedtke, R. J., Brusven, M. A., and Watts, F. J.,** Benthic insect community changes in relation to in-stream alterations of a sediment-polluted stream, *Melanderia,* 23, 21, 1976.
52. **Wesche, T. A.,** Stream channel modifications and reclamation structures to enhance fish habitat, in *The Restoration of Rivers and Streams,* Gore, J. A., Ed., Butterworth, Boston, 1985, chap. 5.
53. **Thompson, C. S.,** Stream relocation on surface mined land, in *Fish and Wildlife Relationships to Mining,* Starnes, L. B., Ed., Water Quality Section, American Fisheries Society, Milwaukee, Wis., 1985, 39.
54. **MacArthur, R. H. and Wilson, E. O.,** The Theory of Island Biogeography, Princeton University Press, N.J., 1967.
55. **Gore, J. A.,** Benthic invertebrate colonization: source distance effects on community compostion, *Hydrobiologia,* 94, 183, 1982.

56. **Minshall, G. W., Andrews, D. A., and Manuel-Faler, C. Y.,** Macroinvertebrate recolonization of the Teton River, Idaho: further evidence for the application of island biogeographic theory to streams, in *Stream Ecology,* Barnes, J. R. and Minshall, G. W., Eds., Plenum Press, New York, 1983, 279.

57. **Sheldon, A. L.,** Colonization dynamics of aquatic insects, in *The Ecology of Aquatic Insects,* Resh, V. H. and Rosenberg, D. M., Eds., Praeger, New York, 1984, 401.

58. **Pratt, J. R. and Cairns, J., Jr.,** Export of species to islands from sources of differing maturity, *Hydrobiologia,* 121, 103, 1985.

59. **Bogatov, V. V.,** Method for calculating migratory activity and drift distance of the benthos of large rivers, *Gidrobiol. Zh.* 21, 86, 1985.

60. **Vaux, W. B.,** Interchange of stream and intergravel water in a salmon spawning riffle, Spec. Sci. Rep. Fish. 405, U.S. Fish and Wildlife Service, Washington, D.C., 1962.

61. **McNeil, W. J. and Ahnell, W. H.,** Success of pink salmon spawning relative to size of spawning bed materials, Spec. Sci. Rep. Fish. 469, U.S. Fish and Wildlife Service, Washington, D.C., 1964.

62. **Wesche, T. A.,** The WRRI trout cover rating method, development, and application, Water Resource Ser. No. 78, University of Wyoming, Laramie, 1980.

63. **Cooper, C. O. and Wesche, T. A.,** Stream channel modification to enhance trout habitat under low flow conditions, Water Research Ser. No. 58, University of Wyoming, Laramie, 1976.

64. **Seehorn, M. E.,** Trout stream improvements commonly used on southeastern national forests, in *Proc. Rocky Mtn. Stream Habitat Management Workshop,* Wiley, R., Ed., Wyoming Game Fish Department, Laramie, 1982.

65. **Duff, D. A.,** Historical perspective of stream habitat improvement in the Rocky Mountain area, in *Proc. Rocky Mtn. Stream Habitat Management Workshop,* Wiley, R., Ed., Wyoming Game Fish Department, Laramie, 1982.

66. **Maughan, O. E., Nelson, K. L., and Ney, J. J.,** Evaluation of Stream Improvement Practices in Southeastern Streams, Bull. 115, Virginia Polytechnic Institute and State University, Blacksburg, 1978.

67. **Gard, R.,** Creation of trout habitat by constructing small dams. *J. Wildl. Manage.,* 52, 384, 1961.

68. **Smith, B. H.,** Not all beaver are bad; or, an ecosystem approach to stream habitat management, with possible software, 15th Annu. Meet. Colorado-Wyoming Chapter, American Fisheries Society, Fort Collins, Colo., 1980.

69. **Winget, R. N.,** Methods for determining successful reclamation of stream ecosystems, in *The Restoration of Rivers and Streams,* Gore, J. A., Ed., Butterworth, Boston, 1985, chap. 6.

70. **Claire, E. W.,** Stream habitat and riparian restoration techniques; guidelines to consider in their use, Workshop for Design of Fish Habitat and Watershed Restoration Projects, County Squire, Ore., 1980.

71. **Alvarado, R.,** Minimum design standards for log weirs, U.S. Forest Service, Pacific N.W. Region, Malheur National Forest, 1978.

72. U.S. Forest Service, Wildlife Habitat Improvement Handbook, FSH 2609.11, U.S. Government Printing Office, Washington, D.C., 1969.

73. **Navarre, R. J.,** A new stream habitat improvement structure in New Mexico, *Trans. Am. Fish. Soc.,* 91, 228, 1962.

Chapter 4

ABANDONED MINES IN ILLINOIS AND NORTH DAKOTA: TOWARD AN UNDERSTANDING OF REVEGETATION PROBLEMS

Claus Grunwald, Louis R. Iverson, and Diane B. Szafoni

TABLE OF CONTENTS

I.   Introduction 40

II.   General Problems of Revegetation 40
        A.   Water 40
        B.   Acidity and Potential Acidity 42
        C.   Fertility and Toxicity 42
        D.   Salinity/Sodicity 43
        E.   Soil Atmosphere 43
        F.   Temperature 44
        G.   Topography and Soil Texture 44

III.   An Illinois Case History: The Longwall Mine District 45
        A.   General Site Description 45
        B.   Study Sites 45
                1.   Spring Valley 46
                2.   Standard 48
                3.   Ladd 48
                4.   Wenona 48
        C.   Illinois Site Assessment 49

IV.   A North Dakota Case History: Western Surface Mines 49
        A.   General Site Description 49
        B.   Study Sites 50
                1.   Velva 50
                2.   Davenport 52
                3.   Fritz 53
                4.   New Salem 53
        C.   North Dakota Site Assessment 53

V.   Comparison Between the Illinois and North Dakota Sites 54

VI.   General Conclusions and Reclamation Suggestions 54
        A.   Erosion Control and Moisture Conservation 54
        B.   Organic Material 55
        C.   Fertilizer 55
        D.   Alteration of pH 55
        E.   Salt Management 55
        F.   Plant Selection 55

Acknowledgments 56

References 56

I. INTRODUCTION

Present laws require that land surfaces altered by mining activity be restored. Lands mined prior to the Federal Surface Mining and Reclamation Act of 1977 or to earlier state laws were exempt. Since no reclamation laws were enacted in Illinois until 1962[1] and in North Dakota until 1969,[2] the deep-mine area of the Illinois Longwall District, mined between 1875 and 1930, and the North Dakota surface mines, mined before 1965, were exempt. The 1977 federal act, however, placed a per-ton tax on coal production with the proceeds to be used in the reclamation of abandoned mines. The Illinois and North Dakota abandoned mine sites discussed in this paper were involved in reclamation studies sponsored under this law.

North Dakota and Illinois rank nationally first and third, respectively,[1] in total identified coal resources. North Dakota has the largest lignite deposits and Illinois the largest bituminous deposits. Approximately 65% of Illinois overlies coal-bearing strata of the Eastern Interior Coal Field, with a total coal resource of 162 billion tons;[3] North Dakota has an estimated 351 billion tons of lignite reserves beneath the western half of the state in the Northern Great Plains Region.[4]

Over 41,700 ha of land in Illinois were affected by surface and deep mining prior to the reclamation law of 1962.[1] When the entire Eastern Interior Coal Region is considered, 66,750 ha not previously covered by law are currently in need of reclamation by any law.[5] About 100 abandoned mines ranging in size from 0.1 to 12 ha are found in the Longwall mining area along the Illinois River in north-central Illinois (Figure 1). Even after more than 50 years, most of the sites remain barren and from all indications will not revegetate by natural means. Some spoil piles are almost 60 m high, and many of these steep, barren slopes continue to erode. The deposited sediment causes flooding and water quality problems that affect adjacent farm land.

The North Dakota coal region, located in the western part of the state (Figure 1), is presently mined entirely by surface mining techniques. Earlier underground operations had been generally disbanded in the 1940s because of shallow, unconsolidated overburden. Reclamation following coal extraction was not generally practiced until the 1969 surface mine reclamation law was enacted by the North Dakota legislature.[6] The North Dakota regulatory authority, the Public Service Commission, recognizes 616 abandoned mine sites in the state; over 350 of these were surface mines widely distributed across the western half of the state. These sites range in size from <0.5 to 730 ha and are equally varied in condition of spoil and in vegetative cover.[6]

The purpose of this chapter is threefold: (1) to review the problems of revegetation in general terms; (2) to describe in detail four Illinois and four North Dakota sites, including both abandoned deep-mined and surface-mined areas that have shown little or no natural revegetation; and (3) to draw several conclusions about the problems of and the possibilities for revegetating these areas.

II. GENERAL PROBLEMS OF REVEGETATION

The growth and survival of plants depend on a number of environmental factors, such as moisture, nutrients, temperature, and light. In disturbed areas, such as mined areas, additional man-introduced factors (i.e., changes in soil pH, texture, salinity or sodicity, or alteration in existing conditions such as changes in availability of soil moisture, soil stability, or soil temperature) influence plant survival. The environmental variables discussed here generally have the greatest impact on reestablishing vegetation in abandoned mined areas.

A. Water
Water is crucial for plants, particularly during seed germination since the water content

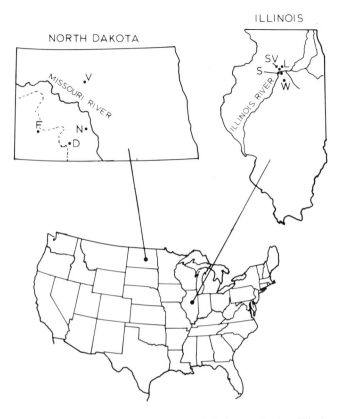

FIGURE 1. Location of Illinois and North Dakota study sites. Illinois sites: Spring Valley (SV), Standard (S), Ladd (L), and Wenona (W). Dakota sites: Velva (V), Davenport (D), Fritz (F), and New Salem (N). The dotted line in the North Dakota map shows the line of glaciation.

of dormant seeds may be as low as 5%. Before germination can take place, seeds must imbibe relatively large quantities of water.[7] Because this water is derived mainly from the soil, the ability of soils to retain water is a critical factor in the revegetation of barren mines.[8] The availability of water to a plant depends upon the amount held per unit volume of soil and upon the depth the roots of the plant can penetrate. Soil texture, therefore, is an important variable. Sandy soils hold less water than fine textured soils, and the addition of organic matter greatly improves the water holding capacity of soils, particularly of sandy soils or mine spoils. Unfortunately, this treatment is often very expensive.

The water status of soils and plants can be defined in terms of water potential. The amount of water available to a plant depends upon the relationship between the water potential of the soil and the water potential of the plant.[9] As long as the water potential of the soil is higher (less negative) than that of the plant, the root system of the plant will absorb moisture. However, when the soil moisture drops below the water potential of the plant, the plant wilts. If no water is added, permanent wilting results. While no precise value can be given since a number of factors including plant species are involved, the value of -1.5 MPa is generally cited as the permanent wilting point.[10]

As soil loses moisture from its pores, water transport diminishes. Soil texture, therefore, is important in determining water conductance. In addition, shrinkage may occur, both in the soil and in the roots of plants, thus decreasing contact between the roots and the soil.[11] Due to coarse particle size and lack of soil colloids, rapid changes in surface soil moisture have been observed at a number of mine sites,[12,13] and newly germinated seedlings and nursery stock may be unable to cope with the dryness of the soil surface during initial establishment. After plants are established, soil moisture conditions often improve.

Irrigation is generally too costly to be part of any revegetation effort. However, the application of mulch can be used to increase soil moisture, decrease evaporation, and improve growing conditions.

B. Acidity and Potential Acidity

In general, plants grow at optimum rates under neutral soil conditions. Unfortunately, much of the abandoned mine wastes are either quite acidic (Illinois) or alkaline (North Dakota).

However, the actual pH of the soil as well as its potential acidity must be considered in determining soil condition. Soil pH can have both direct and indirect effects on plants.[14,15] Direct effects occur at relatively high H^+ or OH^- ion concentrations. At pH values below about 4, H^+ cause a decrease in availability of such nutrients as P, K, Mg, and Ca,[16-18] and these effects are even more damaging at high temperatures,[19] which are often encountered in barren mine spoils.

Indirect effects of H^+ or OH^- can occur at more moderate pH values, and the following dysfunctions can be encountered in acidic soils:

1. Impaired absorption by plants of P, Ca, Mg, and K. A deficiency of P can become a severe problem because of inherently low P levels in mine waste.
2. Increased availability of Al, Mn, and sometimes Fe, Cu, Zn, and Ni. The uptake of heavy metals may result in toxicity.
3. Creation of unfavorable biotic conditions, e.g., impaired N fixation, especially below pH 6.0; inadequate mycorrhizal activity, resulting in reduced absorption of P and K and increased infection by some soil pathogens, particularly fungal pathogens.

The indirect effects of soil alkalinity commonly include:

1. Impaired soil release of Fe, Mn, B, P, Cu, and Zn.
2. Increased availability of Mg, Ca, S, and K.
3. Increased infection by some soil pathogens, especially the Actinomycetes and bacteria.

In general, most abandoned mine sites in the eastern U.S. are acidic. The acidity of mine waste is nearly always attributable to pyrite, which upon weathering produces sulfuric acid. Liming is the most common approach in reducing soil acidity; however, unless all of the pyrite has oxidized, the acidity problem cannot be permanently corrected without repeated applications of lime. When the pyrite has oxidized, the mine waste will eventually lose its acidity as the sulfuric acid is leached by rain.

Adjusting the pH of alkaline mine sites is more difficult. Applications of sulfur or even sulfuric acid have been used to lower the pH, but success has been limited.

C. Fertility and Toxicity

Chemical fertilization can be decisive in the success of revegetating efforts, particularly since such physical factors as moisture and temperature are difficult to control. Most mine waste is low in N and P,[20] and revegetating generally requires the application of nutrients at frequent intervals until sufficient organic matter has accumulated. The buildup of fertility may require several years; however, the addition of N and P alone does not always correct the nutrient deficiency. The addition of mulch usually improves the nutritional value, the infiltration of moisture, and the leaching of salts and toxic ions.[21]

The leaching of nutrients can be a major problem when water infiltration is rapid, when little organic matter or clay is present in the soil to retain ions. Prolonged leaching of mine spoils, especially at low pH, can result in loss of available macronutrients (Ca, Mg, K) and micronutrients (B, Cu), thereby limiting the establishment of plants and reducing their growth.

In acid soils, Al toxicity is often manifested as a P deficiency. Excess Al may reduce the solubility, uptake, and utilization of P.[22] Adding P may detoxify Al, increase P uptake, and prevent P deficiency. Mn is an essential element; however, the toxic levels often found in mine waste may cause the top of plants to yellow and become stunted.[23] If the mine spoil is alkaline, Fe deficiency (plant chlorosis) may occur, especially in the presence of high Ca concentrations.[24]

The cation exchange capacity of soil also gives a measure of its nutrient status, particularly levels of N, P, and K. The exchange reaction is generally carried out with ammonium ions. In neutral soils, the exchange capacity is constant and suitable for plant growth. In both acidic and calcareous (alkaline) soils, however, other reactions may interfere with the exchange process.

D. Salinity/Sodicity

The electrical conductance (EC) of the soil solution is a measure of the water soluble salt content or salinity of the soil. Salts at high levels inhibit plant growth. Very low readings, however, may indicate nutrient deficiency.[25] Salt-tolerant plant species grow in soils with readings above 3.6 mmhos/cm (in 1:1 extracts), but plant growth in general may be reduced at values as low as 1.8 mmhos/cm. High salt levels may adversely affect plant metabolism and/or membrane permeability;[26] however, the relative importance of excess salt or of the water deficit induced by osmotic stress cannot always be established with confidence. Tolerant plant species either cope with high internal ion concentrations or avoid ion uptake and adjust cell turgor by synthesizing organic acids.

Salinity can be a problem in the reclamation of abandoned mine sites, particularly where the water table is near the surface and evaporation rates are high. Salt buildup can also occur in areas of low precipitation, where leaching of the soil is low or where the accumulation of salts is near the surface. When salinity is a problem, effective drainage must be provided or the leached salts will accumulate in low lying areas or in lower soil profiles. When water accumulates, so does salinity.[27]

Sodic soils are those with an exchangeable sodium percentage (ESP) >15 with or without appreciable salinity. Soils with ESP >15 usually have enough exchangeable Na to interfere with the growth of most plants, and the pH is usually alkaline. As ESP increases, soil particles become dispersed and are therefore less permeable to water. As a result, soil exhibits poor structural stability.

The sodium absorption ratio (SAR) is a measure of both the concentration and the composition of the salts in solution in the soil and gives a good indication of the likelihood of successful plant establishment. The SAR is the ratio of Na to Ca and Mg, and soils with values exceeding 12 are unlikely to support plant growth.[28] Salinity and sodicity problems frequently develop in mine overburdens throughout the Great Plains.

E. Soil Atmosphere

Variables such as bulk density, water content, and gas content should be considered when soil is evaluated. The soil matrix determines to a large degree how water moves into and through the soil profile.[29,30] When the soil is saturated or water logged, the pore space is filled with water; when the soil is very dry, most of the pore space is filled with gases. Gases in the soil exist in three states: the free state, which fills the empty pores; the water-dissolved state; and the matrix-adsorbed state.[31] For optimum plant growth, the matrix pores require a particular ratio of gases and moisture. The important gases are oxygen (O_2) and carbon dioxide (CO_2), and concentration gradients develop when roots and soil organisms consume O_2 and produce CO_2. When soils are flooded, O_2 supplies are reduced or absent, and respiration is impaired. In addition, the normal exchange of gases from the roots of plants to the aqueous soil medium is frequently impeded.[32]

F. Temperature

The further temperature deviates from the general biological norm of 20°C, the more important it becomes as a determinant in the geographical distribution of plants.[33] During the growing period of plants, the soil (root) temperatures are usually lower than the air (leaf) temperatures. Variations in the temperature of the root zone are less than those of the ambient air (leaf zone). Indeed, the roots of most plants would be killed if exposed to the variations and durations of temperatures to which their leaves are subjected. Soil temperature is affected by many factors, including (1) air temperature, (2) intensity, quality, and duration of radiant energy, (3) precipitation and evaporation, (4) soil color and thermal conductivity, (5) topography, and (6) surface cover.[34]

The easiest of these six factors to control following mining is surface cover. The beneficial effects of mulch in reducing soil temperature and improving soil moisture is well recognized.[35] The establishment of vegetation also has a beneficial effect on soil temperature. The surface temperature of the soil is especially affected because plants absorb heat from the soil and lose it through the evaporation of water from the leaves and transpiration through air convection over the leaves. Thus, the leaves of plants are kept within a few degrees of the air temperature. Barren spoils, especially the dark shales common on mine sites, lose much less heat through convection and water evaporation than soils covered with vegetation. As the barren soil surface dries, evaporation decreases and a rise in soil surface temperature occurs. Temperatures as high as 60°C have been recorded on mine waste.[25]

G. Topography and Soil Texture

The waste material from many abandoned mines is concentrated in large piles with steep slopes, with resulting instability and excessive erosion. An alteration in topography during mine reclamation, therefore, can have a very positive effect on soil temperature, and this change can be particularly important when seeds are germinating and when seedlings are becoming established. Sun-facing slopes receive more radiant energy, warm more quickly in the spring, and are generally more favorable for plant growth than north-facing slopes. During the summer, however, the soil on sun-facing slopes may dry out more quickly.[36] As a result, the seeding of barren mine sites must be timed to ensure maximum benefit of seasonal moisture. In addition, plant species should be chosen according to their suitability for north- or sun-facing tolerance.

Most abandoned mines have spoil materials with textures and structures quite different from those of normal soils. Spoil materials generally contain little organic matter and have little or no microbial activity. Depending upon the mining procedure used and the removal of waste material, the site can be quite homogenous or relatively heterogenous. The use of heavy earth-moving equipment during reclamation can also cause considerable compaction, thereby increasing bulk density and decreasing soil porosity and water infiltration.[25] Changes in bulk density affect the rate of water and air movement in the soil, as well as the diffusion of ions. Initially, ion diffusion increases with increased soil bulk density, but beyond a certain point the diffusion rate decreases rapidly as bulk density increases.[37] Since compact soils have a limited amount of moisture, they offer a less hospitable environment for plants. Gas exchange at the root zone may also be limited, further inhibiting plant growth.[38] Finally, large root tips have greater difficulty penetrating compact soil than do the tips of small roots.

FIGURE 2.   A typical example of the mine spoils in the Longwall Mine District. The site
is located near Standard, Ill.

III. AN ILLINOIS CASE HISTORY: THE LONGWALL MINE DISTRICT

A. General Site Description

The historic Longwall Mine District in Illinois, the only place in the nation where large
tonnages of coal were mined by the Longwall method, is located at the northern edge of an
extensive coal-bearing stratum that covers approximately 65% of the state.[1] Its mines generally lie within 50 km of the Illinois River (Figure 1). Longwall mines are circular and
have the overall appearance of a wheel. The hub represents either the pillar of coal left
unmined around the hoisting and ventilation shaft or an area packed with rock after the coal
has been removed. The haulageways throughout the mine represent the spokes of the wheel,
and the rim is the working face. This type of mine operated in Illinois from about 1875
through 1930, well before any reclamation laws were passed, and most of the over 100
refuse piles from these mines remain barren today. A typical example of the mine spoils in
the Longwall District is shown in Figure 2.

The land of Longwall District is mainly used as cropland today. It was once covered by
Wisconsinan glaciation, which resulted in a topography that is generally level to rolling.[39]
The major stream valleys and the extensive Shelbyville and Bloomington moraines provide
the greatest relief.[40] Mesic, black-soil prairies, marshes, and prairie potholes in the young,
poorly drained glacial drift were once characteristic of this region.[41]

The climate of northern Illinois is continental with hot, humid summers and cold winters.
The average mean temperature variation is from 10° to 12°C. Precipitation during the growing
season of 160 to 170 days is fairly uniform with an average annual precipitation of 80 to
85 cm.[42] A climate diagram for Peru, a town near the center of the study area, shows the
average annual fluctuation in moisture and temperature over a 20-year period (Figure 3).
As can be seen, high temperatures in late summer, coupled with decreasing precipitation,
result in stressful growing conditions.

B. Study Sites

The principal coal mined in the Longwall District, the Colchester (No. 2) seam, was a

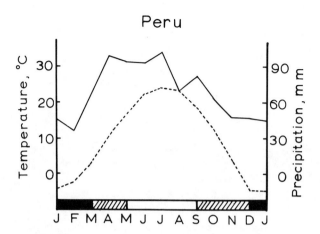

FIGURE 3.   A 20-year average climatic diagram for Peru, Ill.
which has an elevation of 187 m, a mean annual temperature of
10.6°C, and a yearly precipitation of 856 mm. The solid line is
mean monthly precipitation and the broken line is mean monthly
temperature. Months with mean temperatures below 0°C are shaded;
months with one or more days below 0°C are indicated by slanting
lines. The frost-free period is left blank.

member of the Carbondale Formation of the Kewanee Group of the Pennsylvanian System.
Due to the unique mining technique and the narrow (about 1 m) width of the coal seam, a
large amount of roof rock was also removed to gain access to the coal. Thus, the large mine
waste piles at the Longwall sites were composed of the overlying Francis Creek shale from
the roof rock and the underlying underclay, with additions of coal scraps. These waste piles
were flammable and often burned, producing several types of waste.[43]

One type of waste was gray in color and was the least altered. It was not burned and has
basically retained its original color. Although this material was composed largely of shale
and clay, it also included pyritic coal, sandstone, cut-off ends of timbers, and a small amount
of other refuse discarded by the miners. The second type of mine waste, altered by low-
temperature oxidation, was soft and reddish-brown in color. It contained a few partially
burned fragments that were hard and had sharp edges. The third type (clinker material) was
waste that had burned to a brick-hard cohesive mass. Usually vivid shades of red, it tended
to occur as masses of sharp fragments.[43]

The occurrence of the mine waste was essentially uniform across the study area. The gray
waste was more acidic than the other types, probably because it was the least altered by
burning. The EC was also higher for this material. All types of waste had similar levels of
Al, but the gray spoils had the highest levels of Fe and Mg. The N and P levels were highest
in the oxidized red wastes.

The vegetation of 11 of the more than 100 waste sites was inventoried in 1982 as part of
a study for the Illinois Abandoned Mined Lands Reclamation Council.[44] The spoil characteristics of four of these sites (Figure 1), Spring Valley, Standard, Ladd, and Wenona, were
studied intensively during the summer of 1982. Additional inventories of vegetation and
further spoils studies were carried out during the spring of 1983.

1. Spring Valley

The Spring Valley site (SV in Figure 1) is located near the western edge of Spring Valley
in Bureau County. The mine opened in 1886 and closed in 1947. At one time this site
consisted of two mine piles. The larger of the two was almost entirely removed years ago,
and the area it occupied has partially revegetated. The smaller pile remains and was essentially

Table 1
SELECTED PARAMETERS AS DETERMINED BY FIELD
SAMPLING ON FOUR ILLINOIS SITES

	Spring Valley	Standard	Ladd	Wenona
Vegetation				
Cover, %	<5.0	<5.0	20.0	90.0
Total number of species	42.0	57.0	41.0	70.0
Perennial/annual + biennial ratio	1.6	3.9	3.9	2.0
Spoil				
pH[a]	3.33	3.70	3.80	4.95
Lime requirement[b]	8.65	7.13	7.58	4.68
EC, mmhos/cm[a]	2.40	1.53	0.85	0.13
N, ppm, total[c]	905.0	597.0	952.0	2360.0
P, ppm[d]	0.8	4.0	7.4	2.2
K, ppm[d]	418.0	446.0	874.0	1581.0
Ca, ppm[d]	1013.0	1202.0	787.0	1507.0
Mg, ppm[d]	574.0	326.0	151.0	264.0
Al, ppm[d]	942.0	531.0	756.0	485.0
Fe, ppm[d]	73.9	132.6	56.9	5.9
Si, ppm[d]	43.7	57.3	73.2	106.9
Zn, ppm[d]	4.8	3.5	3.7	20.9

Note: Because analytical methods differed between the Illinois and the North Dakota studies, comparisons between states should be made with caution.

[a] In 1:1 soil water extracts.
[b] Lime requirement in tons per 1000 tons of waste.
[c] Total Kjeldahl N.
[d] Dilute acid (0.05 HCl + 0.05N H_2SO_4) extractable analyzed by direct reading emission spectrometry.

barren. Its steep slopes were severely eroded with extensive gullies, and was essentially barren. In an attempt to contain the mine waste, a ditch had been dug to the south of the pile. Runoff from the pile could be seen along the road to the north and in the wooded area to the west. In more recent times, removal of material from the eastern side of this pile has resulted in the loss of some adjacent vegetation.

Sparse vegetation was found on the lower slopes of the smaller pile. The dominant species were sour dock (*Rumex acetosella*), barnyard grass (*Echinochloa pungens*), and common smartweed (*Polygonum pensylvanicum* var. *laevigatum*). A small patch of woody plants covered the northern side of the pile, mainly boxelder (*Acer negundo*), elm (*Ulmus* spp.), and oaks (*Quercus macrocarpa* and *Q. velutina*). The ratio of perennials to annuals + biennials was very low (1.6), a finding characteristic of early successional species (Table 1). Average vegetation coverage was <5%.

Of the four Illinois mine sites discussed here, the mine waste at the smaller Spring Valley pile had the lowest soil moisture potential during 1982. The southeast and southwest sides of the pile had soil moisture potentials below −2.5 MPa during much of the summer. Due to the steep slopes, the water status of the soil did not improve during rains; as a result, revegetation by natural means appears unlikely. This site also had the highest Al (942 ppm) and salinity (EC 2.4 mmhos/cm in 1:1 extracts) and the lowest pH (3.3), findings that probably reflect a low leaching rate (Table 1). Of the four Illinois sites, Spring Valley would require the highest application of lime to bring the spoil to pH 6.5. The combination of low pH, toxic or near toxic levels of Al, high lime requirement, excessively dry slopes, and somewhat elevated salinity makes Spring Valley the most inhospitable to plant growth of the Illinois sites.

2. Standard

The Standard site (S in Figure 1), located in Putnam County, consisted of two large mine waste piles just outside a residential area.[45] One of the piles was extensively burned and the other less so. Both piles, however, were extensively eroded and had affected neighboring farm fields and a nearby drainage ditch.

The piles were essentially barren with sparse patches of grasses herbs, and trees. Dominant ground cover included sour dock, pale dock (*Rumex altissimus*), common ragweed (*Ambrosia artemisiifolia*), and heath aster (*Aster pilosis*). The scattered tree growth was dominated by large-toothed aspen (*Populus grandidentata*) and boxelder. Extensive seeps were present on the northwest and southeast slopes and had a thick growth of bryophytes, grasses, and herbs. The water eventually drained out onto an extremely saline mudflat on which spear scale (*Artiplex patula*) was the only plant species found. The perennial to annual + biennial ratio was 3.9 (Table 1). Vegetation cover was <5%, with most of the plant cover on the less steep slopes.

The pH of the waste material averaged 3.7 and the lime requirement was 7.13 tons/1000 tons of waste (Table 1). Total N level was very low (597 ppm), but none of the heavy metals appeared at toxic levels. During the summer months, the soil moisture of the southern slopes was very low, generally between -1.5 and -2.0 MPa.

3. Ladd

Located in Bureau County, the Ladd site (L in Figure 1) consisted of two extensively burned mine waste piles. The piles were barren and extensively eroded. One pile had been planted with black locust (*Robinia pseudoacacia*) in the mid-1970s. Many of the trees still survived on the southern side of the pile.

Dominant tree cover was the planted black locust. Other species present included boxelder and cottonwood (*Populus deltoides*). Prominant ground cover species included bladder campion (*Silene cucubalus*), sour dock, pale dock, and common ragweed. The perennial to annual + biennial ratio was the same as for the Standard site (3.9), but because the Ladd site had been planted, vegetation cover was about 20% (Table 1).

The mine waste had a pH of 3.80 and a rather high lime requirement at 7.58 tons/1000 tons of spoil (Table 1). Only Al at 756 ppm appeared to be at toxic or near toxic levels. The moisture level of the soil fluctuated greatly during the growing season, particularly on the southeastern and southwestern sides of the piles, but seldom did it drop below -1.5 MPa. The fluctuation in soil moisture appeared to reflect rainfall patterns.

4. Wenona

The Wenona site (W in Figure 1) in Marshall County is located in an urban residential area. Roads confined three sides and a railroad bordered the fourth. The top of the pile was removed by the Army Corps of Engineers in the 1950s. The site was extensively modified for revegetation and species planted were black locust, honeysuckle (*Lonicera* spp.), and smooth brome (*Bromus inermis*).

The site supported full ground cover and moderate overstory vegetation, with >90% overall coverage. The ratio of perennials to annuals + biennials, however, was rather low at 2.0 (Table 1). The dominant ground cover was smooth brome and honeysuckle. The overstory species included black locust and slippery elm (*Ulmus rubra*). The Wenona site had the highest pH (4.95) and the lowest lime requirement (4.68 tons/1000 tons of waste) (Table 1). The level of organic matter was fair and the N level was good. This site had the lowest salinity (EC 0.13 mmhos/cm) and Al level (485 ppm), findings that probably reflect a relatively high leaching rate. The soil moisture level was adequate during most of the summer, but some water shortage was observed during late September.

C. Illinois Site Assessment

The naturally revegetated areas of the mine waste piles at the four Illinois sites were found primarily around the perimeter of the piles, on slumps, and in gullies. In all cases, areas that had revegetated naturally had slopes of <30°. Nearby steeper slopes were barren due to more rapid runoff and to the tendency of the water to wash seeds to adjacent flatter areas. Water often accumulated on these flatter areas, providing moist places for germination. Areas with seeps also exhibited increased revegetation due to additional water, as long as salinity was not excessive.

Most of the plant species were "weedy" in nature (Table 1), which produced easily dispersed seeds. Species generally reflected nearby seed sources. Other studies[46] indicate that the seeds of most invading species are dispersed by birds or by the wind. Due to the harsh conditions of the Illinois mine waste piles, however, few plant species were able to survive.

The spoils of the two barren sites (Spring Valley and Standard) had a lower pH, a higher lime requirement, and a higher salinity than the partially revegetated Ladd site and the more completely revegetated Wenona site (Table 1). Although the high levels of salinity found at Spring Valley could not be tolerated by salt-intolerant plant species,[25] salinity did not generally create a special problem, with the exception of local areas surrounding seeps. Mg was higher in the spoils from the barren mine sites; conversely, Ca, Zn, K, and N were higher in the spoils from the revegetated site. Of the heavy metals, only Al and Zn were found at highly elevated levels. Elevated Al and Zn can be tolerated by many plant species, especially if the soil pH is adjusted. N was well below the minimum requirement for plant growth at the barren and partially vegetated sites (Table 1). P, another major plant nutrient, was low at all sites including the vegetated site at Wenona; in the presence of high Al, further induced P deficiencies can occur. In general, application of fertilizers would greatly stimulate plant growth.

Among the parameters investigated on the Longwall mine waste, the most limiting factor for plant growth was insufficient water.[44] The barren site at Spring Valley consistently had the lowest available moisture. Wenona, the vegetated site, conserved its moisture. When it rained, the mine waste of the barren sites became "very wet" (saturated), but the rate of drying was very fast (Figure 4). Available moisture on the barren sites fluctuated greatly because the moisture-holding capacity of the mine waste was highly similar to that of sand (Figure 4). In addition, the slopes of the barren piles are very steep, many exceeding 35°. As a result, rainfall ends up as runoff, accentuating the problem of inadequate moisture. The steep slopes of the piles also facilitate erosion, often undercutting or burying establishing seedlings. The establishment of vegetation is difficult under present conditions; however, reducing the steepness of the slopes would permit the penetration of rainfall and reduce erosion.

IV. A NORTH DAKOTA CASE HISTORY: WESTERN SURFACE MINES

A. General Site Description

The western half of North Dakota is largely underlain by lignite coal. This coal is presently extracted by a few large surface mining operations. Historically it was mined by many small operations, both surface and underground.

Land use in western North Dakota at present is nearly equally divided between grazing/hayland and cropland. Except for the extreme southwestern corner of the state, much of the area was glaciated during Wisconsinan times; however, the depth of the glacial drift decreases

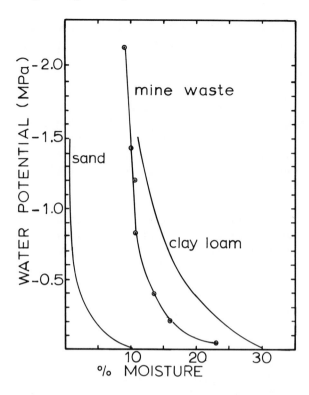

FIGURE 4.   Moisture-release curve of mine waste from the Standard site. For comparison, curves of typical sand and clay loam are also given.

near the southern boundary and little drift exists southwest of the Missouri River (Figure 1). Deep till overburdens like those found in the northern part of western North Dakota are generally much more hospitable to plant growth than the Paleocene shales and sandstones characteristic of the region south of the Missouri River.

The Paleocene sediments were deposited about 65 million years ago on a flat, sometimes swampy plain, similar to parts of the modern day coastal plains of the southeastern U.S. Sands, silts, clays, and lignite were deposited, depending on the fluvial nature of the locality.[47] This material usually is composed of nutrient deficient sands or Na rich, sodic montmorillonite.

The climate of western North Dakota is semiarid and continental, characterized by extremes in temperature and dry periods, especially in late summer (Figure 5). Annual rainfall averages only about 44 cm, and the moisture deficit is the major stress factor.

B. Study Sites

During 1982, 19 abandoned mine sites were intensely sampled and analyzed for numerous variables critical to plant establishment.[48] Four of these sites — Velva, Davenport, Fritz, and New Salem — are discussed in this paper. A typical site is shown in Figure 6.

1.Velva

Velva (V in Figure 1), in the northern part of North Dakota, is located in the glaciated region on the Missouri Coteau and had fairly deep glacial deposits. The mine spoils were steep (>30° slope) but had revegetated naturally to a large degree. The cover was relatively diverse. The presence of water impoundments and wooded draws supported good vegetative cover. The mine was still active in 1982. After over 50 years of mining, it was about 500

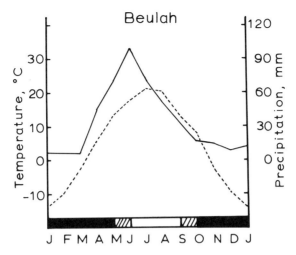

FIGURE 5. A 25-year average climatic diagram for Beulah, N.D., which has an elevation of 543 m, a mean annual temperature of 4.7°C, and a yearly precipitation of 437 mm. The solid line is mean monthly precipitation and the broken line is the mean monthly temperature. Months with mean temperatures below 0°C are shaded; months with one or more days below 0°C are indicated by slanting lines. The frost-free period is left blank.

FIGURE 6. Abandoned mine area at the Fritz site near Bowman, N.D.

ha in size and had been the subject of a detailed study in natural succession.[49] The overburden material was dominantly glacial till with a texture of sandy clay loam, a pH about 7.5, low salinity, low SAR, low Na, low-to-moderate P, and fairly high N (Table 2). With the exception of a slight P deficiency, these spoils showed no overwhelming revegetation problems.

The dominant vegetation on the Velva sites was yellow sweet clover (*Melilotus officinalis*); however, many other species were present, including white prairie aster (*Aster ericoides*),

Table 2
SELECTED PARAMETERS AS DETERMINED BY FIELD
SAMPLING ON FOUR NORTH DAKOTA SITES

	Velva	Davenport	Fritz	New Salem
Vegetation				
Cover, %	73.0	76.0	9.0	3.0
Species diversity[a]	10.3	9.5	3.3	2.8
Total number of species	25.0	25.0	6.0	5.0
Perennial/annual ratio	5.8	1.3	1.2	0.3
Spoil				
pH	7.50	7.19	3.77	8.13
EC, mmhos/cm[b]	1.1	0.9	2.7	9.7
Saturation, %	39.0	44.0	47.0	104.0
Bulk density, g/cm^3	1.01	1.08	1.28	1.13
Organic matter, %[c]	1.54	0.51	0.4	0.30
N, ppm[d]	14.0	18.0	3.0	2.0
P, ppm[e]	2.8	25.3	53.0	1.5
K, ppm[b]	41.5	30.1	26.2	63.7
Ca, ppm[b]	254.1	197.8	70.5	128.1
Mg, ppm[b]	45.2	42.6	28.9	240.9
Na, ppm[b]	26.0	12.7	506.7	2508.2
Na adsorption ratio	0.04	0.2	7.2	32.4
Mn[f]	3.3	2.1	4.7	3.7
Fe[f]	9.5	13.1	80.7	6.4
Zn[f]	0.38	0.47	0.45	0.49
Cu[f]	0.59	0.28	0.59	0.57

Note: Because analytical methods differed between the North Dakota and the Illinois sites, comparisons between states should be made with caution.

[a]   Shannon index of diversity.
[b]   In saturation extracts.
[c]   Walkley-Black carbon.
[d]   Available N as ascertained by anaerobic incubation.
[e]   0.5 N Na bicarbonate extractable P.
[f]   Diethylenetriaminepentaacidic acid extractable.

smooth blue aster (*A. laevis*), field bindweed (*Convolvulus arvensis*), stiff goldenrod (*Solidago rigida*), quack grass (*Agropyron repens*), western wheatgrass (*A. smithii*), and Kentucky bluegrass (*Poa pratensis*). Average cover was 73%, species diversity was 10.3, and the ratio of perennials to annuals + biennials was 5.8 (Table 2). The relatively high proportion of perennials indicated dominance by later successional species.

2. Davenport

The Davenport site (D in Figure 1) had revegetated naturally with many trees and herbaceous cover. Because the site was located on the southern limit of glacial advance, the amount of till in the overburden was negligible. The mine site was surrounded by pasture and contained steep (>30°) spoil banks, high walls, and a pond.

The spoil was sandy loam with a neutral pH, low salinity and SAR, and a fairly high level of major nutrients. Compared to the spoils on the Velva site, the spoil material at Davenport had a lower level of organic matter, a slightly higher bulk density, and a slightly higher saturation percentage (Table 2).

Melilotus officinalis was the primary plant species at Davenport. Other important plants were prickly lettuce (*Lactuca serriola*), blue wilt lettuce (*L. oblongifolia*), smooth brome

(*Bromus inermis*), Russian thistle (*Salsola* spp.), and blue grama (*Bouteloua gracilis*). Average cover was 76%, and species diversity was 9.5 (Table 2). Although species diversity and richness were essentially the same at Velva and Davenport, Davenport had a higher preponderance of annual, early successional species as indicated by the perennial to annual + biennial ratio of 1.3.

3. Fritz

The Fritz site (F in Figure 1), south of the glacial front, is located within a grazed pasture and is, for the most part, barren (<10% cover). The spoils were not steep (5° to 10°) and had a compact sandy structure.

The spoils had moderate-to-low salinity (EC 2 to 4 mmhos/cm in saturation extracts), relatively high SAR (13 to 16), very low available N (2 to 4 ppm), and very high P (>50 ppm) (Table 2). Zn, Mg, and K were low, while Na and Fe were high. Contrary to general occurrences for western mines, the pH was acidic (pH 3.7) rather than alkaline. The major factors contributing to poor vegetative growth appeared to be low N and K and the poor unfavorable water-holding characteristics of the spoil, i.e., high bulk density and low organic matter.

The site had very low species diversity (3.3) and an even division between perennials and annuals + biennials. The most common species was white sage (*Artemesia ludoviciana*), but several legumes were also present, presumably because of high P and low N levels, a condition that gives nitrogen fixers a distinct advantage. The low pH probably minimized nitrogen fixation, however.

4. New Salem

Situated on the fringe of the glaciated region, the New Salem site (S in Figure 1) is located in a prairie pasture and has been abandoned for at least 20 years. Slopes were moderately steep (20° to 30°) and excessive sheet erosion due to the lack of vegetation and the texture of the material was apparent.

The spoils were very low in available N (1 to 3 ppm), P (1 to 2 ppm), Zn (0.5 ppm), Ca (<128 ppm), and organic matter (0.3% with measurements as low as 0.03%) (Table 2). They were high in salinity (EC about 10 mmhos/cm) and in sodicity (SAR up to 40 and Na of 2500 ppm). The pH (8.1) was also high, further reducing nutrient availability.

Vegetation on the site was poor, with only 3% cover, a 2.8 species diversity, and perennial to annual + biennial ratio of only 0.25. Species that survived were weedy in nature and included gumweed (*Grindelia squarrosa*) and summer cypress (*Kochia scoparia*).

C. North Dakota Site Assessment

Ranked in decreasing order of revegetation difficulty, the four North Dakota sites appear as follows: Velva, Davenport, Fritz, and New Salem. Since all sites had been abandoned for at least 20 years and since all were adjacent to natural seed sources, natural revegetation should have had ample opportunity to occur. Revegetation success, therefore, can be attributed at least in part to the quality of surface spoil material. Glacial till like that found at Velva has good texture and a fair nutrient content. The Davenport site did not have a thick layer of glacial till, but it had revegetated nearly as well as Velva. The addition of fertilizers would have been beneficial. The Fritz site, on the other hand, was very sandy, compacted, and deficient in N. The addition of organic matter and N would have improved revegetation success. The New Salem site was sodic, fine textured, and very low in major nutrients and organic matter. The likelihood of natural revegetation was small, and major alterations would be needed for revegetation to occur.

The major problem at each of the four sites, however, was lack of moisture. The 25-year climatograph for Beulah, a town in the center of the study region, suggests that moisture deficits, as indicated by the intersection of the temperature and precipitation curves, are common in late summer (Figure 5). When coupled with barren slopes and excessive runoff, moisture deficits are exacerbated.

V. COMPARISON BETWEEN THE ILLINOIS AND NORTH DAKOTA SITES

The case histories make clear that revegetation problems in Illinois and North Dakota are in some respects quite similar but in others quite different. The most severe problems in each state are water related. At the North Dakota sites, steep slopes, high Na, and low, late summer rainfall caused extremely dry conditions for part of the growing season. At Illinois sites, even steeper, highly eroding slopes caused equally difficult establishment conditions. The relatively greater rainfall and fewer Na problems in Illinois were negated by the additional sheet erosion, which causes physical displacement of plant propagules. In addition, the naturally occurring plant species adjacent to the Illinois sites were not as well adapted to severely water deficient conditions as were the naturally occurring species near the North Dakota sites.

Major differences between the two regions were pH, salinity, and seed sources. Illinois mine waste seldom had pH levels above 4.0; North Dakota spoils were seldom below pH 7.0. Salinity and sodicity problems were more prevalent in North Dakota, where montmorillonite spoils are common and where rainfall is insufficient to leach salts readily. The natural vegetation types and the current use of the adjacent land also differed between the two regions. North Dakota sites were frequently surrounded by semiarid native prairie and thus had a greater chance of receiving seeds adapted to that stressful environment. In Illinois, most adjacent land was in row crop agriculture; consequently the immigration of seeds appropriate for the colonization of mine spoils was less likely.

The availability of nutrients appeared to be a restriction for plant growth in both North Dakota and Illinois. In Illinois, low pH and high potential acidity created inhospitable conditions. In North Dakota, high pH was generally found; however, under high pH conditions, P is often complexed as Ca phosphates, thus creating severe P deficiencies. Micronutrients like Cu, Fe, or Zn can also be deficient under high pH and excessive under low pH.

The lack of organic matter also contributed to poor plant growth in both Illinois and North Dakota. Organic matter is critical for the retention of soil moisture and also provides a mechanism for the slow release of nutrients. In both regions, the poorest sites had negligible organic material, and that insufficiency can be considered both a cause and an effect of poor vegetation.

VI. GENERAL CONCLUSIONS AND RECLAMATION SUGGESTIONS

Similarities in conditions between the Illinois and North Dakota regions suggest overlapping strategies for reclamation. Several recommendations are outlined below.

A. Erosion Control and Moisture Conservation

Reducing the slopes of the waste piles, especially in Illinois where they often exceeded 35° with lengths greater than 100 m, would stabilize the banks,[43] prevent slumping, and greatly reduce sheet and rill erosion. Long, devegetated slopes, even though less steep, are nevertheless vulnerable to the formation of gullies. The reshaping of mine wastes, however, is a very expensive practice and additional land must usually be used to store the excess material. Reshaping, therefore, must be cautiously implemented and is best carried out in combination with other treatments to reduce runoff and minimize environmental contamination.[50]

Judicious grading can also enhance the collection and retention of moisture. In some cases, especially the large Illinois waste piles, drainage systems would need to be installed to control excessively high water tables, which may lead to slope instability.[43] Grading can be accomplished in a number of ways to allow water infiltration into vegetation root zones.[50,51]

The addition of soil stabilizers and mulches should be utilized where possible. Numerous products have been used effectively, from submerged burlap[52] to chemical stabilizers to wood residues.[53] Mulches, artificial or natural, reduce evaporation and permit infiltration, and therefore greatly enhance the probabilities for plant establishment.[53] Of course, irrigation can hasten establishment of plants in areas particularly deficient in water.

B. Organic Material

The addition of organic material would help revegetation efforts by increasing the retention of moisture as well as the availability of nutrients.[54-56] The addition of sewage sludge has been shown to be highly beneficial in restoring organic C, N, P, and K. Agricultural productivity can be restored on many mined lands. At a site in Fulton County, Illinois, for example, Chicago sludge was incorporated between 1972 and 1979, and the land is currently used as cropland.[57] Other organic materials, such as bark or manure, are also highly effective.[58] If topsoil can be borrowed from adjacent areas, it can restore much of the productivity of a site.[59]

C. Fertilizer

Careful soil sampling and analysis is essential in order to correct nutrient deficiencies and toxicities. The addition of organic materials helps restore nutrients; however, chemical fertilizers are generally also needed. Deficiencies of most soil derived essential nutrients have been reported at most abandoned sites, except for the micronutrients Mo and Cl.[60] In most cases, deficiencies can be corrected by adding the appropriate nutrients, especially if the application is repeated for several years until soil conditions improve.

The use of N-fixing organisms can benefit long term N management.[61-63] Similarly, inoculation of plant seedlings with mycorrhizae increases nutrient uptake, particularly P.[64,65] Remedial fertility maintenance should continue as a normal procedure.

D. Alteration of pH

Spoil pH plays a large role in determining the availability of nutrients.[25] In North Dakota, many sites were alkaline, a condition that could cause deficiencies of P and of certain micronutrients.[24] Sulfur or sulfuric acid have been used to reduce the pH and thereby to improve nutrient availability. This approach may benefit the North Dakota sites. In Illinois, excessive acidity nearly always occurs on deep mine waste piles. Liming helps to bring the pH to tolerable limits; however, care must be taken to incorporate the lime deeply, at least to the depth of root development.[66] Over liming can cause temporary P deficiencies.[67]

E. Salt Management

Salinity was a problem primarily in North Dakota. In Illinois, it was a problem only in local areas surrounding seeps near the pile bases. High salinity and low pH can result in double jeopardy to plants, as has been reported for some bentonite spoils in Montana.[68] Several techniques can reduce salt levels on such soils, including leaching and the incorporation of gypsum.[28]

F. Plant Selection

Selection of plant species suited for growth on abandoned mines is critical to revegetation success. Since moisture deficits often occurred in both North Dakota and Illinois, the use of drought tolerant species is warranted. Particular pH and salinity ranges should be considered when selecting plants as should the nutrient and germination requirements of particular species. Several sources are available to aid in the selection of plants.[69-74]

The techniques suggested above have proven effective in the revegetation of barren sites, and their technology is continually being refined as research and development continue. Many of these techniques, however, are costly, and it is unlikely that elevated expenditures can be justified for more than the most severely damaged sites. Economic benefits can sometimes outweigh the costs,[75] but such inexpensive means as community sponsored tree and shrub plantings of tolerant species are quite effective in reducing the extent of these environmentally damaging mine waste spoils.

ACKNOWLEDGMENTS

This research was supported in part by the Illinois Abandoned Mined Lands Reclamation Council and the North Dakota Public Service Commission.

REFERENCES

1. **Nawrot, J. R., Klimstra, W. D., Jenkusky, S. M., and Hickmann, T. J.,** Illinois state reclamation plans for abandoned mined lands, *Resource Document,* Abandoned Mined Lands Reclamation Council, Springfield, Ill., 1982, 233.
2. **Beck, R. E.,** The North Dakota surface mining control and reclamation law, in *Some Environmental Aspects of Strip Mining in North Dakota,* Wali, M. K., Ed., North Dakota Geological Survey Educational Series 5, Grand Forks, N. D., 1973, 109.
3. **Smith, W. M. and Stall, J. B.,** Coal and water resources for coal conversion in Illinois, Ill. State Water Surv. and Ill. State Geol. Surv. Cooperative Resources Report 4, 1975, 79.
4. U.S. Geological Survey, Mineral and water resources of North Dakota, North Dakota Geol. Surv. Bull. 63, Grand Forks, N. D., 1973.
5. Soil Conservation Service, The status of land disturbed by surface mining in the United States, Basic Statistics by State and County as of July 1, 1977, SCS-TP-158, Washington, D. C., 1979.
6. **Imes, A. and Wali, M. K.,** An ecological-legal assessment of mined land reclamation laws, *N. D. Law Rev.,* 53, 359, 1977.
7. **Larcher, W.,** *Physiological Plant Ecology,* 2nd ed., Springer-Verlag, Berlin, 1980, 303.
8. **Etherington, G. K.,** *Environment and Plant Ecology,* John Wiley & Sons, London, 1975, 347.
9. **Slayter, R O. and Taylor, S. A.,** Terminology in plant and soil-water relations, *Nature (London),* 187, 922, 1960.
10. **Fischer, R. A. and Turner, N. C.,** Plant productivity in the arid and semiarid zones, *Annu. Rev. Plant Physiol.,* 29, 277, 1978.
11. **Huck, M. G., Klepper, B., and Taylor, H. M.,** Diurnal variation in root diameter, *Plant Physiol.,* 45, 529, 1970.
12. **Richardson, J. A. and Greenwood, E. F.,** Soil moisture tension in relation to plant colonization of pit heaps, in *Proc. Univ. Newcastle upon Tyne Philos. Soc.,* 1, 129, 1967.
13. **Thompson, D. N. and Hutnik, R. J.,** Environmental characteristics affecting plant growth on deep-mine coal refuge banks, Department of Environmental Resources on the Commonwealth of Pennsylvania Spec. Res. Rep. SR-88, 1971.
14. **Black, C. A.,** *Soil-Plant Relationships,* 2nd ed., John Wiley & Sons, New York, 1968, 792.
15. **Hewitt, E. J. and Smith, T. A.,** *Plant Mineral Nutrition,* John Wiley & Sons, New York, 1974, 298.
16. **Jacobson, L., Overstreet, R., King, H. M., and Handley, R.,** A study of potassium absorption by barley roots, *Plant Physiol.,* 25, 639, 1950.
17. **Moore, D. P., Overstreet, R., and Jacobson, L.,** Uptake by calcium by excised barley roots, *Plant Physiol.,* 36, 53, 1961.
18. **Moore, D. P., Overstreet, R., and Jacobson, L.,** Uptake of magnesium and its interaction with calcium in excised barley roots, *Plant Physiol.,* 36, 290, 1961.
19. **Jacobson, L., Overstreet, R., Carlson, R. M., and Chastin, J. A.,** The effects of pH and temperature on the absorption of potassium and bromide by barley roots, *Plant Physiol.,* 32, 658, 1957.

20. **Barrett, J., Deutsch, P. C., Ethridge, F. G., Franklin, W. T., Hiel, R. D., McWhorter, D. B., and Younberry, A. D.,** Procedures recommended for overburden and hydrologic studies of surface mines, USDA Forest Service General Technical Report INT-71, Ogden, Utah, 1980.

21. **Grim, E. C. and Hill, R. D.,** Environmental protection in surface mining of coal, EPA-670/w-74-093, Environmental Protection Agency, Cincinnati, 1974.

22. **Foy, C. D. and Brown, J. C.,** Toxic factors in acid soils. II. Differential aluminum tolerance of plant species, *Soil Sci. Soc. Am. Proc.,* 28, 27, 1964.

23. **Hiatt, A. J. and Ragland, J. L.,** Manganese toxicity of burley tobacco, *Agron. J.,* 55, 47, 1963.

24. **Safaya, N. M.,** Delineation of mineral stresses in mine spoils and screening plants for adaptability, in *Ecology and Coal Resource Development,* Wali, M. K., Ed., Pergamon Press, Elmsford, N. Y., 1979.

25. **Bradshaw, A. D. and Chadwick, M. J.,** *The Restoration of Land: the Ecology and Reclamation of Derelect and Degraded Land,* Blackwell Scientific, Oxford, 1980, 317.

26. **Greenway, H. and Munns, R.,** Mechanisms of salt tolerance in nonhalophytes, *Annu. Rev. Plant Physiol.,* 31, 149, 1980.

27. **Kovda, V. A., Van den Berg, C., and Aagen, R. M.,** Irrigation, drainage and salinity, *FAI12INESCO,* London, Hutchinson, 1973.

28. **Richards, L. A.,** Diagnosis and improvement of saline and alkaline soils, U.S. Department of Agriculture Handbook 60, Washington, D. C., 1969, 160.

29. **Currie, J. A.,** Gaseous diffusion in porous media. I. A non-steady state method, *Br. J. Appl. Phys.,* 11, 314, 1960.

30. **Currie, J. A.,** Diffusion within soil microstructure, a structural parameter for soils, *J. Soil Sci.,* 16, 279, 1965.

31. **McLaren, A. D. and Skujins, J.,** The physical environment of microorganisms in soil, in *The Ecology of Soil Bacteria,* Gray, T. R. G. and Parkinson, D., Eds., University of Toronto Press, Canada, 1968, 3.

32. **Feldman, L. J.,** Regulation of root development, *Annu. Rev. Plant Physiol.,* 35, 223, 1984.

33. **Langridge, J. and McWilliam, J. R.,** Heat responses of higher plants, in *Thermobiology,* Rose, A. A., Ed., Academic Press, London, 1967, 231.

34. **Nielson, K. F.,** Roots and root temperature, in *The Plant Root and Its Environment,* Carlson, E. W., Ed., University Press of Virginia, Charlottesville, 1974, 293.

35. **Anderson, D. T. and Russell, G. C.,** Effects of various quantities of straw mulch on the growth and yield of spring and winter wheat, *Can. J. Soil Sci.,* 44, 109, 1964.

36. **Hughes, R.,** Climatic factors in relation to growth and survival of pasture plants, *J. Br. Grassland Soc.,* 20, 263, 1965.

37. **Warncke, D. D. and Barber, S. A.,** Diffusion of Zn in soils. II. The influence of soil bulk density and its interaction with soil moisture, *Soil Sci. Soc. Am. Proc.,* 36, 42, 1971.

38. **Leyton, L. and Rousseau, L. Z.,** Root growth of tree seedlings in relation to aeration, in *The Physiology of Forest Trees,* Thimann, K. V., Ed., Ronald Press, New York, 1957, 467.

39. **Willman, H. B. and Frye, J. C.,** Pleistocene stratigraphy of Illinois, *Ill. State Geol. Surv. Bull.,* 94, 204, 1970.

40. **Leighton, M. M., Ekblaw, G. E., and Horberg, L.,** Physiographic divisions of Illinois, *Geology,* 56, 16, 1948.

41. **Bier, J. A.,** Land forms of Illinois, Ill. State Geol. Surv., (Map), 1956.

42. **Fehrenbacher, J. B., Walker, G. O., and Wascher, H. L.,** Soils of Illinois, *Univ. Ill. Agric. Exp. Stn. Bull.,* 725, 47, 1967.

43. **DuMontelle, P. B., Berggren, B., and Bradford, S. C.,** Geologic studies related to reclamation of historic Longwall Mining Sites, north-central Illinois, Abandoned Mined Lands Reclamation Council, Springfield, Ill., 1983, 300.

44. **Grunwald, C. and Szafoni, D. B.,** Revegetation study of Longwall mine wastes in northern Illinois, Illinois Abandoned Mined Lands Reclamation Council, Springfield, Ill., 1983, 174.

45. **Iverson, L. R., Szafoni, D., and Grunwald, C.,** Factors affecting revegetation of northern Illinois gob piles: a case study at Standard, Illinois, in *Symp. Surface Mining, Hydrology, Sedimentology, and Reclamation,* University of Kentucky, Lexington, 1983, 255.

46. **Ashby, W. C.,** Vegetation development on a strip-mined area in southern Illinois, *Trans. Ill. State Acad. Sci.,* 57, 78, 1964.

47. **Bluemle, J. P.,** Guide to the geology of southwestern North Dakota, N.D. Geol. Surv. Educ. Ser. 9, Grand Forks, N.D., 1980, 37.

48. **Nicholson, S. A., Dancer, W. S., Jordan, J. E. Iverson, L. R., and Nunna, M. K.,** Vegetation-environment relationships of abandoned mines: toward their successful revegetation, stability, and long-term production, Public Service Commission, Bismarck, N.D., 1984, 379.

49. **Wali, M. K. and Pemble, R.,** Ecological studies on the revegetation process of surface coal mined areas in North Dakota. III. Soil and vegetation development of abandoned mines, Report to the Bureau of Mines, Washington, D.C., 1982, 96.

50. **Glover, F., Augustine, M., and Clar, M.,** Grading and shaping for erosion control and rapid vegetative establishment in humid regions, in *Reclamation of Drastically Disturbed Lands,* Schaller, F. W. and Sutton, P., Eds., American Society of Agronomy, Madison, Wis., 1978, 271.

51. **Herricks, E. E., Krzysik, A. J., Szafoni, R. E., and Tazik, D. J.,** The best current practices for fish and wildlife on surface-mined lands in the Eastern Interior Coal Region, Eastern Energy Land Use Team, Office of Biological Services, U.S. Fish and Wildlife Service, Kearneysville, W. Va., 1982, 212.

52. **Heede, B. H.,** Submerged burlap strips aided rehabilitation of disturbed semi-arid sites in Colorado and New Mexico, USDA Forest Service Research Note RM-302, 1975.

53. **Plass, W. T.,** Use of mulches and soil stabilizers for land reclamation in the eastern United States, in *Reclamation of Drastically Disturbed Lands,* Schaller, F. W. and Sutton, P., Eds., American Society of Agronomy, Madison, Wis., 1978, 329.

54. **Sopper, W. E. and Kerr, S. N.,** Revegetating strip-mined land with municipal sewage sludge, EPA-600/52-81-182, Environmental Research Laboratory, Cincinnati, 1981, 7.

55. **Sopper, W. E., Seaker, E. M., and Bastian, R. K., Eds.,** *Land Reclamation and Biomass Production with Municipal Wastewater and Sludge,* Pennsylvania State University Press, University Park, Pa., 1982, 524.

56. **Halderson, J. L. and Zenz, D. R.,** Use of municipal sewage sludge in reclamation of soils, in *Reclamation of Drastically Disturbed Lands,* Schaller, F. W. and Sutton, P., Eds., American Society of Agronomy, Madison, Wis., 1978, 355.

57. **Peterson, J. R., Lue-Hing, C., Gschwind, J., Pietz, R. I., and Zenz, D. R.,** Metropolitan Chicago's Fulton County sludge utilization program, in *Land Reclamation and Biomass Production with Municipal Wastewater and Sludge,* Sopper, W. E., Seaker, E. M., and Bastian, R. K., Eds., Pennsylvania State University Press, University Park, Pa., 1982, 322.

58. **Alderice, L., Howard, R. L., and Grauea, D. H.,** Possible treatments as alternatives to topsoil replacement on surface mine sites, in *Proc. 1981 Symp. Surface Mining, Hydrology, Sedimentation and Reclamation,* Lexington, Ky., 1981, 251.

59. **Drake, L. D.,** Recommendations for rural abandoned mine program (RAMP) in Iowa, *USA Miner. Environ.,* 5, 15, 1983.

60. **Bauer, A., Berg, W. A., and Gould, W. L.,** Correction of nutrient deficiencies and toxicities in strip-mined lands in semi-arid and arid regions, in *Reclamation of Drastically Disturbed Lands,* Schaller, F. W. and Sutton P., Eds., American Society of Agronomy, Madison, Wis., 1978, 451.

61. **Palmer, J. P. and Iverson, L. R.,** Factors affecting nitrogen fixation by white clovers on mine spoils, *J. Appl. Ecol.,* 20, 287, 1983.

62. **Palmer, J. P. and Iverson, L. R.,** Rhizobial nitrogen fixation of mined lands, in *Bridging the Gap Between Science, Regulation, and Surface Mining Operations,* American Society for Surface Mining and Reclamation, Denver, Colo., 1985, 283.

63. **Dawson, J. O., Christensen, T. W., and Timmons, R. G.,** Nodulation by *Frankia* of *Alnus glutinosa* seeded in soil from different topographic positions on an Illinois spoil bank, *The Actinomycetes,* 17, 50, 1983.

64. **Zak, J. C.,** The importance of vesicular-arbuscular mycorrhizae in the reclamation of mine spoils, in *Bridging the Gap Between Science, Regulation, and Surface Mining Operations,* American Society for Surface Mining and Reclamation, Denver, Colo., 1985, 298.

65. **Grossnickle, S. C.,** Ectomycorrhizae: a viable alternative for successful mined land reclamation, in *Bridging the Gap Between Science, Regulation, and Surface Mining Operations,* American Society for Surface Mining and Reclamation, Denver, Colo., 1985, 306.

66. **Fitter, A. H. and Bradshaw, A. D.,** Root penetration of *Lolium perenne* on colliery shale in response to reclamation treatments, *J. Appl. Eol.,* 11, 597, 1974.

67. **Costigan, P. A., Bradshaw, A. D., and Gemmell, R. P.,** The reclamation of acidic colliery spoil. III. Problems associated with the use of high rates of limestone, *J. Appl. Ecol.,* 19, 193, 1982.

68. **Sieg, C. H., Uresk, D. W., and Hansen, R. M.,** Plant-soil relationships on bentonite mine spoils and sagebrush-grassland in the northern high plains, *J. Range Manage.,* 36, 289, 1983.

69. **Cable, D. R.,** Western wheatgrass transplants grow well on raw mine spoil, USDA Forest Service Research Note RM-345, 1977.

70. **Monsen, S. B. and Plummer, A. P.,** Plants and treatment for revegetation of disturbed sites in the intermountain area, in *The Reclamation of Disturbed Arid Lands,* University of New Mexico Press, Albuquerque, 1978, 155.

71. **Redente, E. F., Ogle, P. R., and Horgis, N. E.,** Growing Colorado plants from seed: a state of the art, Western Energy Land Use Team FWX/OBS-8230, USDI Fish and Wildlife Service, Washington, D.C., 1982, 141.

72. **Fulbright T. E., Redente, E. F., and Margis, N. E.,** Growing Colorado plants from seed: a state of the art, Western Energy Land Use Team, FWS/DBS-82-29, USDI Fish and Wildlife Service, Washington, D.C., 1982, 113.

73. **Thornberg, A. A.,** Plant materials for use on surface-mined lands in arid and semi-arid regions, USDA Soil Service, SCS-TP-157, 1982, 88.
74. **Ries, R. E. and DePuit, E. J.,** Perennial grasses for mined land, *J. Soil Water Conserv.,* 39, 26, 1984.
75. **Bernard, J. R.,** Cost-benefit analysis of reclaiming abandoned gob and slurry: west central Illinois. *Reclam. Rev.,* 3, 103, 1980.

Chapter 5

RECLAMATION AND TREATMENT OF CONTAMINATED LAND

Michael A. Smith

TABLE OF CONTENTS

I.   Introduction 62

II.   Background 63
        A.   Sources of Contamination 63
        B.   The Nature of the Problem 64
        C.   The Human Urban Environment 65

III.   Options for Remedial Action 66
        A.   Introduction 66
        B.   Selection of Appropriate Remedial Measures 70
        C.   Long-Term Effectiveness of Remedial Measures 70

IV.   Soil Treatment After Excavation 71
        A.   Introduction 71
        B.   Physical Separation 73
        C.   Solvent Extraction 73
        D.   Chemical Treatment 73
        E.   Microbial Treatment 73
        F.   Heat Treatment 74
        G.   Stabilization/Solidification 74
        H.   Mobile Installations 74

V.   *In Situ* Treatment 75

VI.   Macroencapsulation/Isolation/Containment 76
        A.   Introduction 76
        B.   In-Ground Barriers 77
        C.   Covering Systems 78
                1.   Introduction 78
                2.   Materials 79
                3.   Prevention of Water Ingress 81
                4.   Upward Migration of Contamination 81
                5.   Vegetation 82
                6.   Methane and Other Gases 82

VII.   Monitoring and Evaluation of Performance 82

VIII.   Building on Contaminated Land 84
        A.   Introduction 84
        B.   Worker Safety 85
        C.   Ground Engineering 85
        D.   Services 86

E. Attack on Construction Materials...86
F. Methane and Other Gases 86
G. Combustible Materials..87

Acknowledgments...87

References..87

I. INTRODUCTION

"Contaminated land" can be defined as "land that contains substances that, when present in sufficient quantities or concentrations, are likely to cause harm, directly or indirectly to man, to the environment, or on occasions to other targets." The emphasis in this definition is on the presence of potentially harmful substances and inherent in it is that a "problem" can only be said to exist once a proper site investigation has been carried out and the results evaluated in site-specific terms taking into account, for example, current and planned uses of the site (various other definitions have been proposed[1] aimed generally at restricting the scope to land where "problems" do exist; however, adopting the principles proposed by Holdgate[2] to distinguish between "contamination" and "pollution", this might be better described as "polluted land").

The contamination of land has become of major concern during the past decade in both North America and Europe for three main reasons:

1. Actual and potential risk to human health and safety through, for example, contaminated water supplies or foodstuffs
2. Actual and potential damage to the natural environment
3. Inhibition of the beneficial use of land for agriculture, residential, amenity, or industrial purposes

In the U.S. this interest has arisen primarily from the overt problems caused by the very large number of "uncontrolled" or "abandoned" hazardous waste sites which prompted the CERCLA or "Superfund" legislation and the associated U.S. Environmental Protection Agency (USEPA) activities to identify problem sites and to develop appropriate remedial measures. The primary task to date on most of the so-called "superfund" sites has been to carry out urgent measures to eliminate an on-going problem. While this has on occasion also been the case in Europe and surveys have been carried out in a number of countries to identify potential problem sites (e.g., The Netherlands where over 4000 sites have been identified of which about 1000 are likely to require some form of corrective action), there has been an additional pressure which is the need to "recycle" land (land is the most valuable resource in many countries) previously used for industry or waste disposal for further beneficial use. Often a change of use is involved, e.g., from industry to housing or agriculture, and consequently the possible presence of contaminants potentially harmful, directly or indirectly to humans, has become of concern. Formal policies to deal with such situations have been adopted in the U.K.[1,3,4] and a number of other countries (e.g., Denmark and The Netherlands[5]) and some parts of the U.S. (e.g., California[6] and New Jersey[7]). The

USEPA has been turning its attention to the long-term future of superfund sites and to the problems posed by some forms of old industrial sites.

This discussion is based mainly on the report[8] of the NATO Committee on the Challenges of Modern Society's (CCMS) Pilot Study Group on Contaminated Land published in 1985. This study group brought together technical experts from seven countries (Canada, Denmark, France, West Germany, The Netherlands, the U.K., and the U.S.) to review the technologies available for the long-term restoration of contaminated land. However, this is a fast developing area of technology and some reference has also been made to more recent literature, particularly the proceedings of the International Conference on Contaminated Soil[9] held in Utrecht in 1985. Other useful sources of information are the various handbooks being produced by the USEPA,[10-13] the proceedings of the annual HMCRI conference on uncontrolled hazardous waste sites,[14] and a number of other recent reports and publications.[15-17]

The CCMS Study Group was primarily concerned with the techniques available to either clean-up contaminated soils (essentially to remove the contaminants) or to provide a long-term remedial action while leaving the contaminants in place in the ground. It was restricted to a consideration of "chemical" contaminants present as solids, liquids, or gases and did not concern itself with biological, radioactive, or combustible contaminants.

The techniques described are of equal applicability to the restoration of natural ecosystems and human urbanized environments. General points arising from the discussions of the study group that appear of general importance in both contexts are

1.   The need to consider long-term (e.g., 100 year) effects
2.   The need for monitoring and evaluation of performance, and on occasions, for continuing maintenance
3.   For a multidisciplinary approach, particularly when problems are being defined and solutions considered

The early failure of many restoration schemes on difficult (e.g., chemically polluted) sites is well documented and can be attributed, at least in part, to inadequate prior study and lack of provision for continuing maintenance. There also appears to have been cases where economic and ecological considerations, e.g., a wish to turn sites green at minimum cost and with minimum intervention, have led to inadequate attention being paid to possible adverse effects on those then attracted to the site.

II. BACKGROUND

A. Sources of Contamination

Contamination can be broadly defined as the existence in the ground of hazardous substances at concentrations above "normal" background levels. Substances giving rise to concern include:

- Certain elements and their compounds (e.g., Pb, As, Cd, Hg)
- Organic chemicals
- Oils and tars
- Toxic, explosive, and asphixiant gases
- Combustible materials
- Radioactive materials
- Biologically active materials
- Asbestos and other hazardous minerals
- Corrosive materials

The "concentration" actually present can range from a few milligrams per kilogram trace element to thousands of drums of unidentified organic chemicals stacked on the land surface.

Contamination of the ground itself commonly arises from disposal of wastes, in one form or another, to properly designated landfills, to uncontrolled dumps and tips or to areas within the boundaries of industrial sites. It may also arise from accidental spillages and leakages during plant operation and from the transportation and stockpiling of raw materials, wastes, and finished products. Widespread contamination can arise from the deposition of air- and waterborne emissions. Land can also become contaminated from overdosage with metal contaminated sewage sludge and other similar organic wastes.

Contaminated sites, particularly those of industrial origin, frequently present other problems. They are often "filled" or "made" ground and consequently badly compacted, they often contain massive foundations and underground pipework, tanks, and other structures, and there may be abandoned and derelict, unsafe, and contaminated buildings still standing.

Any former industrial site may be contaminated, but some industrial activities have a high probability of producing contaminated sites.

These include

- Mining and extractive industries
- Smelting and refining (steel works, foundries, etc.)
- Secondary metal recovery (scrap yards, etc.)
- Gas, coke, and tar manufacturing
- Waste disposal
- Wood preservation
- Tanning and associated trades
- Asbestos production and use
- Pesticide and herbicide manufacture or use
- Railways
- Chemical and allied products manufacture
- Explosive and munitions manufacture
- Metal treatment and finishing
- Paint manufacture
- Sewage works and farms
- Oil storage
- Oil production
- Dockland areas
- Acid/alkali plants
- Pharmaceutical, perfumes/cosmetics/toiletries production
- Electronics industry

B. The Nature of the Problem

Whether the contamination found on a particular site will be a problem will depend upon many site-specific factors and how they influence the degree of risk to a set of "targets" arising from the "hazards" presented by the contaminants that are present. Schemes to rate sites in order of priority for action have been drawn up in a number of countries based upon more or less elaborate risk/hazard/target assessments.[5,18-20]

Targets potentially at risk from contaminated sites include:

1. Workers engaged in investigative, remedial, or construction activities
2. Eventual residents (housing) or users (of schools, factories, recreational areas, etc.), including particularly sensitive groups in the population such as small children

3. A wider population owing to pollution of aquifers and water courses, and wind-blown pollution
4. Animals, plants, aquatic life, etc.
5. Building structures, services, and materials, e.g., by corrosion, fire, or explosion
6. Investment if site closure or evacuation is required, or there is major damage to buildings, or a liability suit is brought on the basis of any of the above

Hazards/exposure combinations include the following:

Physical hazards

1. Explosion and fire
2. Subsidence
3. Corrosion of structures
4. Effects on mechanical properties of soil

Toxic hazards to humankind

1. By inhalation
   a. Dust
   b. Toxic gases
   c. Asphyxiant gases

2. By ingestion
   a. From fingers, etc. (particularly children)
   b. Surface contamination of food

3. By indirect ingestion
   a. Uptake of toxicants by edible plants
   b. Contamination of water

4. By contact
   a. Skin irritation, etc.
   b. Absorption through skin

The hazards to humankind may be further categorized as short term, such as those posed to workers by the presence of highly toxic or corrosive substances or flammable gases, or as long term, such as might arise from the presence in garden soils of carcinogens, radioactive materials, or of elements, such as Cd that are taken up by food plants and may accumulate in the body to an undesirable extent over very many years.

C. The Human Urban Environment

People come in contact with soils in the urban environment mainly through play in gardens and amenity areas, through cultivation of their gardens, through consumption of home grown produce, and through recreational use of amenity areas including visual appreciation of vegetation and land form. In this context a healthy ecosystem will be largely man-made and such as to permit us to pursue our various activities without risk to health or safety.

The urban ecosystem might also be taken to include all those artifacts that enable modern society to exist: housing, commerce, industry, and transportation systems. Thus, restoration of a healthy urban ecosystem could be said to include enabling contaminated sites; and already developed or "used" land in general, to be recycled — for the same or a different use. This puts particularly stringent requirements on the methods of reclamation that need to be employed (see discussion to follow on "selection of remedial options" and on "building on contaminated sites").

The human urban environment is generally polluted to a greater or lesser extent, the main determinants being the extent of industrialization, the age of the urban area, the state of economic development of the country as a whole, and the balance between private and public transportation systems. Soils in older urban areas are generally contaminated, particularly with elements such as Pb, Cd, and Zn, due to deposition from the atmosphere, and from break-down of lead paints and galvanizing and the disposal of ashes, soot, and other domestic residues in gardens. Sometimes concentrations are sufficiently high to give rise to concern for the health of residents or to hinder plant growth. In addition, efforts to conserve energy leading to lower ventilation rates in housing, and use of new furnishing materials, are combining to give cause for concern for "indoor air quality". The problems posed by atmospheric pollution have long been recognized and in all industrialized countries steps are being taken to reduce pollution from this source. However, there will always remain a legacy of contaminated urban soils so that whenever such an area is redeveloped, contamination is likely to be found, even in the absence of a history of use for industry or deposition of wastes.

While it is unlikely to ever be possible to clean-up all urban soils this is no reason for not doing so when the opportunity presents itself, due to a redevelopment scheme for example. Nor is it an excuse for applying less stringent criteria in an urban setting than would be applied in an area where pollution is less endemic. The fact that this legacy of contaminated soil is likely to remain indefinitely into the future suggests a need for viable plant communities specific to these environments.

III. OPTIONS FOR REMEDIAL ACTION

A. Introduction

Actions taken in response to the identification of a contamination problem may, in the first instance, be designed to provide a temporary amelioration of an immediate hazard rather than to control pollution of the environment on a longer term basis, e.g., to limit the amount of pollutant reaching a water course rather than to totally prevent the pollutant escaping from the site. However, in all but a minority of cases, the aim should be to find some permanent or ultimate solution that will either eliminate the identified risk or at least reduce it to an acceptable level.

Environmental pollution is usually viewed in terms of a source and pathways to a target (Figure 1). Once a risk is identified the options available are to eliminate the source, remove the target, or cut or restrict the pathways. All three options are available, on their own or together, for dealing with contaminated land (Figure 2).

In the context of recycling land, the "targets" may be removed by altering the use of the land, e.g., while Zn might be of concern if plants are to be grown it is unlikely to be a problem if the land is to be covered with concrete for car parking, factory development, etc. This principle may be applied on a macroscale in zoning land for permitted forms of development or within individual developments.[21]

The most direct and obvious solution in many cases appears to be to remove the offending materials, deposits, and contaminated ground from the site for disposal or treatment elsewhere (i.e., to remove the source). Unfortunately this is not always practicable, nor is it always desirable, nor is it necessarily a permanent solution in the broadest sense.[22,23] The difficulties surrounding excavation are summarized in Table 1 and the aspects of the process are illustrated in Figures 3 to 5.

Excavation and transport of contaminated material may be hazardous, present other en-

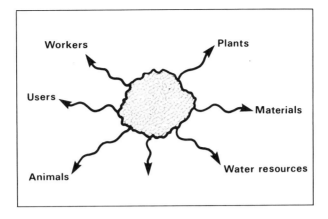

FIGURE 1.   Contaminant source and potential targets.

FIGURE 2.   Basic options for dealing with contamination.

vironmental hazards (e.g., nuisance of traffic movements), and are costly. It may not be possible to find a suitable disposal site for the contaminated material within an economic distance either because of its toxicity or its bulk. Further, redeposition may simply move the problem elsewhere to be rediscovered at some later date. Transportation of material off site for treatment by incineration or chemical means can present similar problems. There are also likely to be community objections to being the recipients of somebody else's waste. Thus there are often strong environmental, economic, and social reasons for dealing with contamination where it is found.

Most of the options available for treatment of contaminated land consist of, or rely on, actions of an engineering type. In considering options it is therefore important to learn the essential lessons that can be drawn from a consideration of the generality of civil and structural engineering works. These are

1.   A recognition that the properties of materials and components used in a structure may deteriorate with time to such an extent that the structure has a foreseeable and limited life
2.   Consequent upon this, recognition of a need to monitor performance and to carry out maintenance to ensure continued performance to acceptable standards
3.   A need to design for natural and man-made interventions of a catastrophic type, e.g., 1 in 100 year flood level

Table 1
DIFFICULTIES OF EXCAVATION

Complicating factors
  May be no sharp edge to contamination
  Contamination may have moved out of the site, including beneath neighboring buildings (Figure 4)
  Excavation may cause hydrogeological problems (e.g., removal of overburden may cause artesian entry of water)
  May need to shore up excavation
  Difficulty in finding disposal site
  Difficulty in finding "clean" fill materials
  Environmental problems — traffic movement, noise, atmospheric and surface water pollution, odor problems
Possible additional actions required
  Underpin adjacent buildings
  Collect, treat, and dispose of contaminated water
  Selectively remove areas of especially high contamination
  Temporarily lower groundwater levels
  Protect rivers, streams, etc.
  Monitor working environment and neighborhood
  Have chemical specialists on site

From Smith, M. A., in *Reclaiming Contaminated Land*, Cairney, T. C., Ed., Blackie, Glasgow, 1986, 114. With permission.

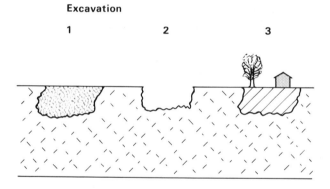

FIGURE 3.   The concept of excavation and replacement with clean fill.

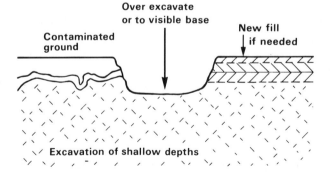

FIGURE 4.   Excavation and replacement to shallow depths — excavation should extend into clean ground and the fill should be placed in thin layers using standard civil engineering procedures.

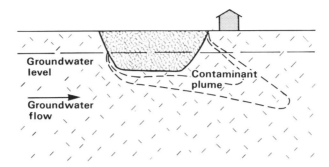

FIGURE 5. The complications of excavation (and other forms of treatment) — adjacent buildings and movement of contaminants out of the immediate area of the site.

Three main methods of treatment were identified by the CCMS Study Group:

1. On-site treatment — defined as being a method of decontaminating or otherwise reducing the environmental impact of the bulk of contaminated materials on a site by excavation, treatment to detoxify, neutralize, stabilize or fix; and usually reposition on site
2. *In situ* treatment — defined as the treatment without excavation of the bulk material on a contaminated site
3. Macroencapsulation — in which a site is contained within a ''box'' usually consisting of a cover, a peripheral barrier, and sometimes a base (obviously very difficult to install if not already present in the form of a natural stratum of low permeability)

The concept of on-site treatment is based on the idea of having fairly mobile equipment that can be moved from site to site as the need arises. However, experience in The Netherlands suggests that this is unlikely to be possible in the European context due to economic and environmental factors and that the provision of a variety of centralized soil treatment facilities is more likely to be viable. The economics may be different in North America where transport distances will be generally greater. However, the environmental and administrative problems are likely to be similar.

The liquid phase within and on contaminated sites is of great importance. It is the usual means by which the contamination moves out into the wider environment and a treatment method should not be adopted without first considering how to deal with the liquid phase. This may be in the form of a neat undiluted contaminant, a strongly contaminated aqueous phase permeating or leaching from the soil, or at the other extreme, a very lightly contaminated groundwater remote from the site. For example:

1. Excavation may be required below groundwater level — groundwater must then be pumped, treated, and disposed of
2. *In situ* treatment by leaching with water or other solvent will produce a contaminated waste stream

Methods for the treatment of heavily contaminated leachates, surface waters, etc., are well established and continuously being improved upon. Such methods were therefore not included in the CCMS Study and are not discussed here. However, the Study looked at the liquid phase in three respects:

1.   The use of hydraulic measures as an adjunct to other treatment options
2.   The treatment of contaminated groundwater involving extraction, treatment, and rein-
      jection or by *in situ* means (such treatment will often be required in association with
      other remedial measures particularly those involving containment of a site)
3.   The modeling of groundwater movement (an understanding of the probable conse-
      quences of containment and other hydrological control measures is essential to their
      design)

B. The Selection of Appropriate Remedial Measures

Various authors[10,23-25] have discussed the selection of appropriate remedial measures generally emphasizing the need for a systematic approach.

Sites differ greatly in size and the nature and extent of contamination; they may cover
several hundred hectares[26,27] or even be regional in extent, but the typical urban site usually
covers only a few hectares or less. They will often be

1.   An old pit or quarry filled with poorly consolidated refuse, rubble, and undefined
      wastes, and will often contain large amounts of combustible materials permeated with
      methane and carbon dioxide.
2.   An old industrial site. The buildings will often have been demolished and the rubble
      and wastes, and hence contaminants, spread indiscriminately about the site in order
      to provide an attractive flat site; there will be old foundations, drains, and buried tanks;
      there will be areas of fill and made ground; the ground will be polluted to a considerable
      depth due to spillages and deliberate dumping of wastes with a range of contaminants
      including asbestos, Pb, Cd, etc., and mobile organic liquids such as oils, tars, and
      solvents; mobile contaminants may have moved outside the site (Figure 4).

Factors to be considered when deciding on what remedial actions are needed on sites that
are to be built upon are listed in Table 2. Those cited in the USEPA Handbook on Remedial
Action[10] include: performance (effectiveness and useful life), reliability, implementability,
safety, institutional and public health concerns, environmental concerns, and costs.

C. Long-Term Effectiveness of Remedial Measures

The concept of long-term effectiveness and the extent to which the various treatment
options available can provide long-term solutions, have been reviewed by Stief.[28]

If the contaminants are destroyed *in situ* or removed then, provided the work is properly
carried out, a permanent and fully effective solution is achieved. If, however, the contaminants are not removed or destroyed, any remedial system adopted will be intended to prevent
or impede the contaminants from reaching the targets at risk. If the treatment is to be
considered permanent, then it must be designed accordingly. This is of course very difficult
to do given present knowledge and experience and indeed the CCMS study group concluded
that: "Very few of the technologies reviewed have been sufficiently proved in applications
specific to the treatment of contaminated land although they may have been tried for other
purposes. In practice, therefore, all technologies, other than those in which the contaminant
is destroyed or rendered harmless, offer a solution of only limited or uncertain duration,
unless other mechanisms, such as microbial attack, reduce the contamination."

It is important to recognize that the term *effectiveness* can be used to mean different things
during the stages that any remedial measure will go through. It can be applied to the
performance of a component part of a remedial system (e.g., cut-off wall) or to the system
as a whole (e.g., cut-off wall plus groundwater pumping). It is necessary to distinguish
between the *theoretical* effectiveness and *installed* effectiveness. In addition, *effectiveness*
can be looked at as effectiveness on an arbitrary scale at a point in time or as the ability of

Table 2
**FACTORS TO BE CONSIDERED WHEN DECIDING ON REMEDIAL ACTION
ON URBAN CONTAMINATED SITE**

1. Present and intended topography and the relation of site levels to surrounding areas, roads, services, etc.
2. Adjacent land areas (e.g., proximity of buildings)
3. Surface drainage, adjacent water sources, groundwater levels and movement, underlying aquifers
4. Propensity of the site for flooding, etc.
5. Location of existing services
6. Maximum depth of excavation required for services or foundations (major services, especially sewers, usually have to be installed at considerable depths and this inevitably means digging into the contaminated material even if all the other works can be kept within clean material)
7. The consequences of settlement within any imposed cover and of settlement in the ground due to imposed loads from the cover or building
8. Safety of workers and neighbors during site works
9. Environmental impact of site works
10. The significance of a future pollution incident on the site
11. Interaction between building works (e.g., foundations and services) and any completed reclamation works
12. The significance of any future site works, e.g., extensions to buildings, repairs to services, change of land use or redevelopment

From Smith, M. A., in *Reclaiming Contaminated Land,* Cairney, T. C., Ed., Blackie, Glasgow, 1986, 114. With permission.

the system to continue to perform to an acceptable standard over a prolonged period of time (*performance* may be a better term for this latter concept).

When considering the probable long-term effectiveness of a proposed remedial measure the following should be taken into account:

1. There will always be flaws in execution and the theoretical effectiveness will never be achieved, i.e., the installed effectiveness will always be less than the theoretical effectiveness.
2. Environmental factors will generally act to reduce effectiveness with time.
3. Intervention by man, including change of land use in subsequent years, may reduce effectiveness.
4. Contamination will not generally reduce with time (methane is an exception).

It is essential, therefore, for the time element to be taken into accurate account in the design of reclamation works. Wherever possible a "design life" should be defined. It is interesting to note that while those responsible for disposal of radioactive wastes require consideration of time spans of hundreds or even thousands of years, requirements regarding toxic wastes are either nonexistent or limited to a few decades at most.

IV. SOIL TREATMENT AFTER EXCAVATION

A. Introduction

The CCMS study group considered soil treatment methods in terms of excavation and treatment of the soil on-site followed by its reemplacement in the excavation. However, as already mentioned, the immediate course of development in The Netherlands, where most effort has been devoted to the subject to date, has been to set up central processing facilities. Similarly, it is important to point out that development has been rapid since the CCMS state-of-the-art review was prepared (April 1984) and while it remains one of the most comprehensive reviews of the theoretically available treatment methods, it is already a little out-of-date in terms of what has been achieved to date. Other sources include the proceedings[9] of the conference held in Utrecht in 1985 and the USEPA Handbook.[10]

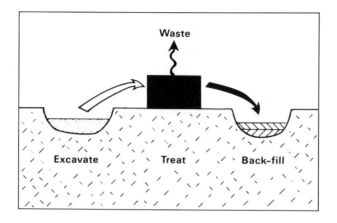

FIGURE 6.   The concept of soil clean-up.

Table 3
SOIL TREATMENT OPTIONS

Physical
    Solvent extraction
    Separation based on
        Gravity
        Particle sizing
        Settling velocity
        Magnetic properties
        Flotation
Chemical
    Neutralization
    Chemical oxidation
    Chemical reduction
    Hydrolysis
    Electrolysis
    Ozonation
    Photolysis
    Enzymatic degradation
Thermal
    Evaporation or destruction through:
        Direct heating
        Indirect heating, including steam stripping
Microbial
Stabilization/solidification
    Cement-based systems
    Polymer-based systems

    The clean-up option for soils is very important because it offers an ultimate solution although there will often be an associated residual waste stream. Once treated either on-site or at a central treatment facility, the soil may be returned from whence it came (Figure 6), disposed of as a waste, or used as fill or for some other beneficial purpose. How it may be used will depend on a variety of factors, including economics and how clean it really is. One of the less fortunate side effects of some treatment systems is that the soil is rendered unsuitable for sustaining vegetation without amendment with fertilizers and organic matter. Soil treatment can be based on physical, thermal, chemical, microbial, or stabilization/solidification processes, and these are listed in detail in Table 3. Many are based on the application of technologies already developed for mineral processing or treatment of wastes and, in some cases, equipment designed for other purposes[29] has been adopted to provide a rapid and economical solution. A number of limitations have been recognized regarding some of the treatment methods. Sandy soils are generally easier to treat by physical means than clayey or organic (peaty) soils. Physical approaches are also generally more suitable for soils containing a single contaminant rather than a range of contaminants. Some still leave a substantial volume of contaminated material requiring disposal or further treatment. Many of the processes require a preliminary preparation step in order to separate the coarsest material (stones, bricks, bits of wood, and concrete), and some crushing, grinding, and homogenization may be required. This is not necessarily an easy process if the material is impregnated with tars or oils, for example. Any segregated coarse material must be separately treated or disposed of. Care must be taken throughout all the treatment methods described below to avoid unacceptable emissions to atmosphere through evaporation or dust blow. Large, protected stocking areas may be required. In addition, as the soil must first be excavated, all the practical and environmental difficulties attendant on this will have to be dealt with, as will those associated with transportation if the treatment is carried off site.

The conceptually available treatment methods have been reviewed in detail by Rulkens et al.[30] (in the CCMS report) and particular forms of treatment (e.g., chemical, microbial) by a number of other authors[31-33] and there are numerous descriptions of particular experimental and full scale applications.

B. Physical Separation

Physical treatment methods take advantage of the tendency for contaminants to be concentrated in a particular size fraction (e.g., metals may adsorb onto the very fine clay particles) or component of a soil (e.g., the organic matter). Various processes of flotation, filtration, settlement, and segregation can be employed to separate out the different size fractions.

C. Solvent Extraction

In solvent extraction the contaminants are dissolved in an extraction agent; this may be water, an aqueous solution, or a nonpolar solvent. The critical step is to obtain intimate contact between the contaminants and the extracting agent and then to physically separate the solid from the liquid phase. This may be comparatively simple with a sandy soil but may be difficult to achieve economically with a clayey soil.

D. Chemical Treatment

The need for intimate mixing of soil and treatment agent (and the subsequent separation difficulties) also arises with the various forms of chemical treatment listed in Table 2. Where the chemical treatment reduces solubility (e.g., reduction of Cr^{6+} to Cr^{3+}) so that the contaminant is precipitated, then a size fractionating step may become an essential feature of the process. General problems include a frequent need for long reaction times and the need to over-dose with the treatment chemical, which might itself be potentially polluting, so that a secondary clean-up step must be incorporated in the overall process.

E. Microbial Treatment

Intimate mixing of soil and treatment agent is also required in microbial treatment. Both microbes and nutrients must be applied to the soil either by mixing in reactors or rotary drums, or by applying agricultural techniques of harrowing or ploughing to treatment beds or during the construction of windrows. Usually the aim is for the microbe(s) to metabolize the contaminants or otherwise act on them to destroy them or at least reduce their toxicity. However, they can also be used to increase the solubility of a contaminant (this applies in particular to metallic contaminants) so that they can then be removed by solvent extraction.

For most compounds, the most rapid and complete degradation occurs aerobically, although there are some, such as lower molecular weight halogenated hydrocarbons, that at present can only be degraded anaerobically.[10]

Microbial treatment can be based either on microorganisms specially bred in the laboratory to deal with particular contaminants or on assemblages of indigenous microorganisms taken from the contaminated site to be treated and cultured in the laboratory before reseeding. While the former may work well when the contamination is fairly simple or particularly resistant the latter approach seems likely to be more fruitful when a range of contaminants is present. The physical and scientific principles underlying microbial treatment have been reviewed by Rees and others.[34] Microbial treatments appear to be particularly promising in the medium term and are being developed in The Netherlands, the U.K., and the U.S., both for application after excavation and, in the longer term, for *in situ* application.[10]

F. Heat Treatment

Heat can be used either to destroy contaminants or to remove them by evaporation. In both cases, the heat may be applied either by direct heat transfer from a flame or heated air (or other gas including, for example, steam — "steam stripping") or by indirect heat transfer, e.g., by external heating of a rotating drum. In all cases a gas cleaning system will be required and care must be exercised in the selection of heating systems to avoid generation of more toxic substances than were already present (e.g., formation of chlorinated dibenzodins from chlorinated hydrocarbons).

G. Stabilization/Solidification

A large number of stabilization and solidification processes have been developed for the treatment of hazardous wastes but, despite many years of work, they are not generally accepted by the competent authorities as providing a totally safe method of disposal to land. These doubts arise from the nature of the physical and chemical processes involved, various experimental studies, and variation in the quality of application that have resulted in an unsatisfactory end product.[10,35,36] Nevertheless, whatever the shortcomings in particular cases, they can be very useful in making otherwise physically difficult materials (e.g., sludges and silts) more easy to handle particularly as some are comparatively easy to employ on site using standard concrete making and related technologies.

Most of the commercially available processes rely on a reduction of the "(micro)-leachability" through fixation of the water and tend to be based on cementitious systems involving the use of such materials as Portland cement, Na silicate, lime, pozzolans (e.g., fly ash), and hydraulic slags. Systems may also be based, however, on chemical reactions to form insoluble compounds, isolating contaminants from water by adding or forming hydrophobic compounds, and control of pH to minimize solubility.

H. Mobile Installations

As indicated above it was hoped in The Netherlands that soil clean-up installations could be mobile and moved from site to site, but in practice this has not generally been achieved although some of the plants are semimobile having been constructed from containerized modules.[33] Among the factors militating against mobile plants have been the need to deal with environmental issues (e.g., potential atmospheric pollution) each time the plant is resited. This appears to have been a problem with the sophisticated mobile incineration system, developed by the USEPA.[37] Mobile plants have been developed in the U.K. for the treatment of metal mining residues for recovery of metal values.[38]

Table 4
IN-SITU TREATMENT
OPTIONS

Physical
    Deep ploughing
    Solidification
    Leaching/solvent extraction
Chemical
    Neutralization
    Precipitation
    Oxidation
    Reduction
    Fixation
    Ion exchange
    Enzymatic degradation
Other
    Heat treatment to fuse soil
    Electrokinetic techniques
    Ground-freezing

From Smith, M. A., in *Reclaiming
Contaminated Land,* Cairney, T. C.,
Ed., Blackie, Glasgow, 1986, 114.
With permission.

V. *IN SITU* TREATMENT

In situ treatment, in which the contaminated ground is treated to remove, destroy, or fixate the contaminants without excavation, is conceptually the most attractive method of treatment since if offers a permanent solution with minimal disturbance of the ground. It is, however, with the exception of some surface or near surface treatments, the most difficult to achieve in practice. Processes considered suitable for application *in situ* are listed in Table 4. There are a number of potential difficulties:[10,39,40]

1.   It may be difficult to ensure contact between treatment agents and contaminants.
2.   It may be difficult to ensure that the treatment has been effectively applied.
3.   There may be a waste stream requiring disposal or treatment.
4.   The treatment agents may themselves be potential pollutants.
5.   There may be possible adverse interactions with soil materials or some contaminants (e.g., increased mobility).

In situ treatments can be divided into (1) those applied at or near the surface, (2) those requiring injection at depth, and (3) those based on thermal[41] or electrical processes.[39,40]

Examples of surface treatment include amendment of metal (Zn, Cd) contaminated soils with lime to raise the pH or with cationic exchange resins[42] to reduce "availability" of metals, deep ploughing to invert the top meter or so of soil, "soil flushing", and solvent extraction.[43,44]

For treatment at greater depths agents have to be injected under pressure. The techniques employed are similar to those used in grouting for engineering purposes and in addition to the difficulty of ensuring contact with agent and contaminants there is the additional one that pressure injection cannot usually be applied at depths of less than about 2 m without danger of fracturing and ground disturbance. This might be overcome by application of a permanent or temporary cover of clean material.

The aim of *in situ* treatment may be to remove, destroy, or immobilize the contaminants

Macro-encapsulation

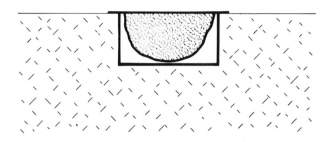

FIGURE 7.   The concept of macroencapsulation/containment.

and consequently the processes listed in Table 4 also appear in Table 3. If microbial treatment is to be used then nutrients must also be provided. These are discussed at length in the USEPA Handbook.[10]

Solidification may be desirable for reasons other than treatment of contamination, e.g., (1) to improve the engineering properties of ground, (2) to stabilize sludge, etc. that presents a physical hazard, and (3) to prevent or stop combustion.

The most obvious agents to use for solidification are cementitious or chemical grouts. Compatibility between grouts and contaminants will clearly be a major consideration.[10]

Some of the agents that are applied are themselves potential pollutants, and therefore *in situ* treatment may only be possible if steps are taken to prevent migration of contaminants and treatment agents from the site, i.e., by creating a reaction vessel within the ground using in-ground cut-off barriers (see below).

More novel methods involve fusion/vitrification[41] of the bulk of the soil using electrodes inserted into the ground and those based on electrokinetic principles (ions and water are induced to move by application of an electric current)[39,40] The latter are used in civil engineering for dewatering soils and have been used to treat alkaline soils. They have been the subject of laboratory studies for possible application to metal contaminated soils and suggested[40] as a possible means of enhancing penetration of grouts and other treatment agents.

VI. MACROENCAPSULATION/ISOLATION/CONTAINMENT

A. Introduction

Macroencapsulation involves retrofitting a containment system around an area of contamination in order to isolate it from the targets identified as at risk. Thus, in its complete form, the contamination is completely boxed in (Figure 7). In practice, however, only partial systems may be employed, or the encapsulation implemented in stages. It is convenient to consider separately in-ground barriers (horizontal and vertical) that are closely allied to other measures that may be taken to control the groundwater regime, and covering systems, which may frequently be applied on their own and are of crucial importance in the restoration of a natural ecosystem. When considering the use of in-ground barriers or containment systems, the probable effects on the groundwater regime should be considered. Computer models and simulations may be useful in this respect.

It is important to recognize that no containment system can be fully secure even at the time of installation and that effectiveness is likely to fall with time: there will always be some leakage, for example, because of flaws in installation or the inherent, if very low, permeability of the materials used to form the barriers. Interactions between the material used and the contaminants may also cause difficulties. For example,

1. Clay with low permeability for water may be more permeable to organic fluids.
2. Chemical reactions between contaminants and clay may increase permeability to water.
3. Contaminants may react chemically with the barrier materials or enhance other degradation process, e.g., corrosion.

B. In-Ground Barriers

The techniques available to install in-ground barriers and the associated hydraulic measures that may be required have been reviewed in detail by Childs[45] and guidance on specific systems has been published by the USEPA.[10,11]

The best in-ground horizontal barrier is probably a naturally occurring one, but when this does not exist then it may be necessary to install one. This is difficult and although techniques mainly based on grouting techniques do exist, they are not well proven.[10,40,45] When they are critical to the scheme, it may be necessary to install two parallel barriers with monitoring points between.

A great variety of techniques has been developed for installing vertical barriers to prevent movement of groundwater as this is often an essential adjunct to civil engineering works but as already mentioned above, there are nevertheless problems or concerns particularly with regard to interaction with various chemicals. The available techniques can be broadly divided into:

1. Those in which the ground is left relatively undisturbed, e.g., driven sheet piles or injected grout curtains
2. Those in which material is excavated or otherwise removed

Although the aim would usually be to install the barrier outside of the previously contaminated zone this may not always be possible so that it must be recognized that any material excavated during the construction of vertical barriers may be contaminated. Maximum feasible depths of application of vertical barriers range from about 5 m (membranes) to 70 m (bentonite-cement "deep-walls"). Some of the barrier systems available are listed in Table 5.

The selection of what form of barrier to use will be governed by such factors as:

1. The degree of integrity required — some leakage may be permissible
2. The depth required
3. How the junction is to be formed with any horizontal barrier — whether this is natural or man-made
4. Resistance to corrosion and other forms of adverse interaction with contaminants
5. Geology and hydrogeology
6. Environmental impact of installation including, for example, noise and vibration generated during the driving of piles
7. Space available for installation and ancillary operations
8. Costs

The finding, in various studies,[10,46-48] that some commonly used barrier systems based on clay were permeable to certain organic chemicals has led to the development of more chemically resistant systems.[46-48]

A low permeability cap may be required as part of the covering system that completes the scheme in order to minimize infiltration into the container that has been formed. Otherwise

Table 5
METHODS AVAILABLE FOR FORMING IN-GROUND BARRIERS

Horizontal barriers
  Chemical grouting
  Claquage grouting
  Jet-grouting

Vertical barriers

Driven and other systems not requiring excavation	Trench and other excavated systems
Sheet steel piling	Clay-filled trench
Vibrated beam slurry wall	Slurry walls
Panel wall	Soil-bentonite slurry walls
Membrane wall	Slurry walls with membrane
	Narrow trench with membrane
	Bentonite-cement "deep-wall"
Jet-grouting[a]	Cast *in situ* pile wall
Grout curtains (cement and chemical)	Precast bentonite-concrete wall
Ground-freezing	Cement-asphalt emulsions
	Chemically resistant bentonites

Combined actions
  Block displacement

[a] Produces waste stream.

it may fill like a bath tub and overflow. It may also be necessary to install a groundwater pumping system to keep the water table within the box low — preferably lower than that outside so that any flow of groundwater occurs inwards. Of course, any water pumped from within the box is likely to require treatment before disposal or reinjection.

C. Covering Systems
1. Introduction

The term *covering systems* was adopted by the CCMS study team in order to emphasize the need for the cover frequently to perform several, often conflicting, functions and in preference to the term *capping* which seems often to be used to imply a single primary function of preventing infiltration of water.[10] This may be just one of the functions of a covering system.

The main function of a covering system is to isolate the contamination from the targets (Figure 8) but it will often have other functions such as to sustain vegetation or perform an engineering role. The system or its component parts may need to control gas and liquid movement; prevent erosion and aid slope stability; inhibit vermin, fires and odors; prevent dust blow; improve appearance; support vegetation; inhibit root penetration; improve engineering properties; and improve resistance to weathering and other natural phenomena.

The complexities of constructing a cover system for landfill sites were described by Lutton et al.[49] and explored further in the context of contaminated land in general by Parry and Bell[50] as part of the CCMS Study. They are recognized in current USEPA guidance.[10]

Factors affecting the effectiveness of a covering system in performing its various functions include:

1. Its ability to control infiltration of water
2. Its ability to prevent upward and lateral migration of contaminants
3. Its ability to bind contaminants physically or chemically
4. Intractions between covering materials, contaminants, and biota
5. Its engineering behavior

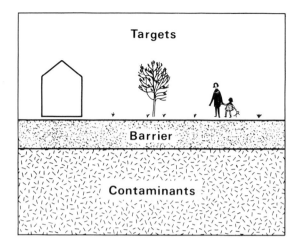

FIGURE 8.   The concept of covering to prevent contact between contaminants and targets.

Because of the many different functional requirements, some of which may conflict, a combination of layers (i.e., a ''cover system'') is more likely to be effective than a single layer of cover. Depending upon circumstances this system might comprise one or more of the following:

1.   Vegetation support layer — top soil, amended subsoil, or waste material
2.   Barrier layer or membrane to prevent passage of liquids or gas (including volatile organics)
3.   Buffer layers to protect barrier layers and to provide intermediate working surfaces
4.   Water drainage layer
5.   Filter layers made of geotextiles or graded aggregates, to prevent breakdown of drainage layers, etc.
6.   Gas drainage layer and associated vents
7.   Break layers to prevent upward migration due to capillary action
8.   Chemical barriers
9.   Barriers to biological processes

Some conceptual systems are shown in Figures 9 and 10.
The material used (see below) must be selected with a view to possible adverse interactions with contaminants and their engineering properties. In addition, the system as a whole must:

1.   Be capable of effective installation, taking into account normal civil engineering tolerances and standards of workmanship
2.   Remain effective for the design life, i.e., be durable
3.   Be resistant to deliberate or accidental intervention by humankind, and to foreseeable natural events such as flooding, climatological extremes, and earth tremors, i.e., be robust
4.   Take account of likely settlement in the cover and in the underlying ground
5.   Require little continuing maintenance

2. Materials
Depending upon the objectives of a particular reclamation scheme, a barrier system will be constructed from one or more layers of (1) natural materials, (2) modified soils, (3) waste materials, and (4) synthetic materials.

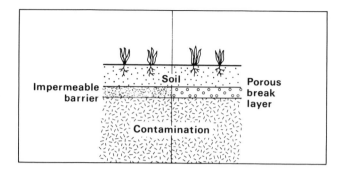

FIGURE 9. Two theoretical ways of preventing upward migration of moisture (and thereby contaminants).

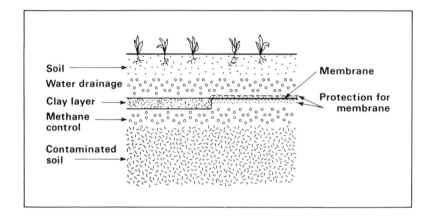

FIGURE 10. A complex covering system intended to control upward and downward migration of moisture and to provide a channel for lateral movement of gas (means to prevent/assist vertical movement of gas will also be required).

Economy will generally favor the use of naturally occurring soils — these being selected on the basis of their physical, chemical, and nutrient status, depending upon what particular role they are required to perform. The permeabilities and engineering properties of soils can be substantially improved by the incorporation of relatively small amounts of materials such as Portland cement, lime, fly ash, and certain other chemicals. This is a widely used engineering procedure in road construction and similar purposes.

Waste materials that can be used, alone or with various admixtures, include fly ash, slags from iron and steelmaking, nonferrous slags, domestic refuse incineration ash, dredged silts, sewage sludge, colliery wastes, and construction rubble. Some of these materials have a potential to cause pollution or may exhibit physical or chemical instability. Their use in a particular case, therefore, requires careful consideration.

Mention has already been made of the efforts to develop very low permeability, clay-based materials for use in forming vertical barriers, and some of these may have a use as a part of a covering system. Flexible membranes have been used to cover very large sites, but they have to be selected and installed with care in order to prevent physical damage during installation and after (e.g., by burrowing animals or tree roots). A wide range of geotextiles are now available, including felts made from polypropylene and other fibers and meshes of various sizes. The fibrous mats may be used to improve the bearing capacity of soft ground (e.g., laid on mud and covered with gravel or crushed rock) as filter layers and to inhibit root penetration. The meshes may be used to inhibit root penetration, to stabilize the surface of slopes, and to reinforce embankments. Synthetic membranes can provide design lives of 20 to 100 years depending upon site conditions and what other components go to make up the covering system.[10]

3. Prevention of Water Ingress

The prevention of water ingress seems generally to have been considered as the main function of covers to hazardous waste sites, but this approach does not seem to take adequate account of those climatic conditions in which there is a net upward movement of moisture (see below). Water ingress can be controlled by the use of natural or synthetic low permeability layers with site drainage and appropriate topography (due allowance must be made for settlement). It is essential when clay-based layers are used that they are kept moist, otherwise cracking to considerable depths may occur. These cracks will not usually heal properly, and more porous soil materials may be washed into them. They must also be protected from freeze-thaw action.

4. Upward Migration of Contamination

There are three main mechanisms that may lead to upward migration of contamination:

1. Uptake by vegetation — subsequent death and decay will lead to a build up in litter and surface layers
2. Cyclic movement in groundwater levels
3. Upward moisture movement due to capillary action

It has also been suggested that worms might cause physical displacement of contaminated material.

Movement due to plants can be limited by using shallow rooted plants, selecting plants with a low tendency to take up contaminants, preventing root penetration into contaminated material, and by preventing upward moisture movement.

When physical movement of water occurs, the main concern regarding contaminants will be for their movement in solution, but this may not be the only mechanism involved. Nonsoluble organic substances may form emulsions, or if less dense than water, float and collect on the surfaces of any free water present in the ground. When they are more dense than water, they may collect on and flow along the surface of an underlying low permeability stratum. Movement may occur against the general direction of groundwater movement. Groundwater levels will tend to show seasonable variations. In the U.K., for example, groundwater levels tend to be highest in February to April and at their lowest in September to October. Groundwater levels can also vary in sympathy with nearby bodies of water. When these are tidal, groundwater levels may show sympathetic variations in level. If selective adsorption of contaminants occurs onto soil particles, fluctuating groundwater levels could lead to a concentration of contaminants in one particular horizon.

Some soil types can give rise to upward capillary movement of moisture in excess of 3 m. Upward movement of water (and other liquids) due to capillary action can be prevented either by means of a low permeability layer or by means of a capillary break layer. The latter is the preferred method unless the impermeable layer is formed from a synthetic membrane. This is because when fine ground materials such as clays dry out they have very large soil suctions and tend to crack (the latter is of equal importance when the aim is to prevent infiltration).[51]

Cairney[51] has discussed the factors governing capillary moisture movement in any particular situation, the way in which it may be predicted from measurement of basic soil parameters, and how to optimize the design of a covering system to minimize possible upward movements using available materials. He described a number of cases where his model has been applied.

5. Vegetation

Vegetation will form a major component of many covering systems as well as being the ultimate goal of many schemes. It can be important because of:

1. Increased evapotranspiration
2. Reduced wind and water erosion
3. Improved appearance
4. As a means of limiting site access
5. Consequential maintenance requirements

Evapotranspiration may be important in reducing the amount of infiltrating water but may cause difficulties by promoting the rate of upward movement of moisture.

The choice of type of vegetation is important in terms of its ability to grow under the particular conditions, its ability to resist any toxins present, and its interaction with biota, e.g., propensity to take up toxic elements that may then be consumed by grazing animals or humankind. The depth of rooting will be important both in respect of soil stabilization and possible adverse effects on the integrity of underlying parts of the system. Judicious choice of densely growing and inhospitable plants may be useful in inhibiting access by humans or animals that might cause damage to surface layers or be affected by any residual contamination. The USEPA has provided guidance[10] on the design of vegetative covers.

While top soil may be a preferred growth medium in many cases, techniques have been widely developed to enable the rapid establishment of vegetation on a whole range of initially inhospitable substrates, and this should rarely be a limiting factor.[52]

6. Methane and Other Gases

The production of methane, carbon dioxide, and other gases is inherent in the disposal of domestic refuse and other organic matter to land. These gases present considerable hazard to man due to risk of explosion, toxicity, and asphixiation, and to vegetation by creating anoxic (oxygen free) conditions in the ground or because of specific toxic effects. Their presence makes it essential to install appropriate methods to control dispersal from the site. A large body of literature on how this may be done now exists.[10,53,54]

VII. MONITORING AND EVALUATION OF PERFORMANCE

It is an accepted requirement in many fields of engineering that some form of monitoring may be required to give an assurance of safety and to ensure the longevity of the structure (e.g., bridges and dams). It is also considered to be an essential means of checking design assumptions and to gather performance data on which to base cheaper and more robust designs in the future. Modern automated instrumental techniques are used to gather data. In addition, computer simulations are widely used to predict behavior. Similarly monitoring may be an essential component of a reclamation scheme and can range[28,55] from a simple visual inspection carried out at regular intervals to the installation of permanent instrumentation (e.g., monitoring wells). The CCMS study group drew a distinction between monitoring to check whether a remedial system was working as planned (e.g., detecting any early signs of failure) and evaluating performance. The latter was seen as essentially a scientific study to see how the system was working or why it was failing and generally requiring a much more detailed appraisal. In practice there is little experience available of such monitoring and evaluation studies and only a few accounts of such studies have been published.[56,57] Stief[28] and Cairney[55] have reviewed the subject of long-term monitoring. Both refer to the need for the design life concept, common in other fields of engineering, to be applied to the design of reclamation schemes, and Cairney[55] also discusses the concept of safety margins.

It is important before embarking on any long-term monitoring program to establish adequate base-line data. Analytical procedures and methods may change unavoidably during the course of an investigation, and it is essential to set aside appropriate control samples for use throughout the study and to retain samples of key materials.

Most reclamation schemes will involve a variety of physical measures such as installation of drainage. These generally have a limited lifetime and are subject to failure for a variety of reasons, e.g., blockage by silt or roots. The functioning of such supplementary measures must be checked from time to time.

The response that is to be made to the results of a monitoring exercise must be considered at the outset and criteria for action set whenever possible. What is to be done if failure or potential failure is indicated? If nothing can be done, or to do so would be technically and/ or economically unfeasible, then the design of the remedial action should be reviewed. It may be that a solution is required that is less susceptible to failure or easier to remedy should it go wrong.

Monitoring must start at the time of installation or before. It is essential that the works are carried out as designed. Otherwise not only may premature loss of effectiveness occur, but the monitoring data may be interpreted incorrectly.

For soils treated after excavation, a need for monitoring and evaluation will only arise when the method of treatment adopted has been stabilization/solidification. Appropriate laboratory tests should have been carried out before application in the field so that, provided the process has been carried out properly, the basic questions will concern the possible consequences of adverse environmental factors (such as infiltration of acid precipitation, freeze-thaw action, and vegetation growth) and the effectiveness of any supplementary actions that have been carried out (e.g., installation of liner or provision of cover). Any evaluation of behavior once the materials have been deposited in the ground will be concerned primarily with (1) the physical and chemical condition of the redeposited material and (2) the possible migration of contaminants into surrounding ground. It should be possible to monitor adequately using permanent well points and by taking cores. However, great care is needed when doing the latter in order to avoid damaging any containment and to avoid letting water into the deposited material.

When *in situ* treatment involves removal or destruction of contaminants, the effectiveness can be checked by appropriate sampling at the time and treatment continued until the design criteria are met. The question of long-term effectiveness does not generally arise. However, the question does arise when the effect of the treatment is to reduce availability to the environment as a result of chemical change, physical change, or change in physical condition of the ground. Chemical reactions are reversible, and most materials used to bring about physical changes are subject to chemical or physical deterioration. The use of lime to increase the pH of a soil and thereby reduce the availability of Zn and Cd to plants is an example of a process of limited duration: lime will be leached progressively from the soil and the pH will tend to fall with time so that repeat applications of lime will be required. Ideally in this case, not only should the soil pH be monitored, but also metal concentrations in appropriate indicator plants. Typical grout materials used to solidify and stabilize soils will also be subject to deterioration due to chemical and physical action.

The effectiveness of in-ground containments in which contamination is not removed or destroyed will generally require sampling of (1) fluids within the site, (2) fluids outside the site, and (3) soil samples from within the site.

In addition, critical environmental factors should be monitored, for example, whether drainage systems are working correctly and the composition of any water entering the site. The monitoring of in-ground barriers calls for the installation of well points within and especially outside the containment. However, this is not as straightforward as it might at first appear. Decisions have to be made about where to install wells, what samples to take, and what to analyze for. Such measures may show that a barrier is or is not working. What they cannot show is the cause of any leakage that is detected.

The methods to be used in the monitoring and evaluation of the performance of a covering system will depend on the functions that the system is required to perform. They range from the simple to the complex, both at the time of installation and in following years. The time dependent changes that may affect performance of the system are major complicating factors. Most covering systems will be multifunctional, and each function may require its own separate monitoring procedure although it may be sufficient to concentrate on one or two key functions. In general, it is not possible to confirm the functioning of a cover system at the time of installation except possibly in respect of those involving gas or water flow. A drainage system might be assessed in terms of flow and chemical analysis, but it would be difficult to be sure of the integrity of a membrane or clay sealing layer when the leakage rate is very low. Functions such as those to sustain vegetation, prevent root penetration, or prevent upward moisture movement cannot be evaluated until several years after installation. In addition, it may be difficult to do so without damaging the system.

The simplest and most essential measure is the systematic and planned visual inspection aided by simple instrumentation such as gas detectors to examine the state of vegetation cover and to note any signs of deterioration such as seepages, erosion, or undesirable animal activity. Proper records, including photographs, should be kept of such inspections. Vegetation can be monitored for uptake of metals or other contaminants. Certain physical characteristics, e.g., water levels and soil suction, should be capable of automatic monitoring, and it should be possible to monitor liquid and gases within and around the site using permanent and temporary installations. However, some factors, such as root penetration or deterioration of materials (e.g., a plastic membrane), are very difficult to monitor without destruction of the system. If it is required that this is done, it is almost certainly better to designate a study area at the outset and to carry out such extra work as may be required to make excavation safe. It may be necessary to rely on indirect evidence from special tests (normal and accelerated) established at the time of installation of the cover system. The fact that changes are gradual is a benefit but will make a judgment of when action is required more difficult. It may be necessary, for example, to monitor metal levels in vegetation over several years to be sure that any observed increase is real. Even then, assessing what it means will be difficult., e.g., does it pose a threat to health or does it portend a further deterioration?

VIII. BUILDING ON CONTAMINATED LAND

A. Introduction

Much of the land in older industrial areas and urban areas may be contaminated (some of our newer industries are also causing environmental problems). While total clean-up may be a desirable goal, the scale of the problem is such that this may not be practicable — nor may it be necessary — although this is a question open to debate. Thus, in order to return land to beneficial use or as a component part of the remedial works, it may be necessary to build on contaminated sites. This need is likely to have a direct effect on the scope of remedial measures that are needed and restrict those that may be employed in a particular case. It is unlikely that remedial measures can be safely decided upon and carried out without reference to the future use of the site. Construction problems in relation to contamination can also arise incidental to some other activity. For example, a tipped site may lie in the path of a new road, pipeline, or sewerage system. Thus, like it or not, it may be necessary to carry out construction works on and in contaminated ground. This is part of restoring the urban ecosystem in the broadest sense.

There are five main reasons for concern regarding contamination and constructional activities:

1. Worker safety at the time of construction
2. Worker safety during any work on the site after completion of the development
3. Health and safety of occupants/users of site
4. Potential for attack on construction materials
5. Adverse environmental effects during and subsequent to development

The question of when a hazard exists and what degree of risk might be acceptable, which arises in any reclamation scheme, is most acute when the intended use of the site is such as to bring users into intimate and regular contact with the site. There is a relationship between end use and the risk to sensitive targets and, hence, also between end use and the acceptable level of contamination and the remedial actions required.

B. Worker Safety

The need to take proper precautions when contaminated sites are being investigated and remedial actions are being carried out is well established. It is recognized that, in certain cases, the extensive protective measures that must be taken can make working very difficult and can lead to physical distress among workers if conditions are not carefully monitored. Such extreme precautions are rarely likely to be acceptable to the average construction worker. The aim in designing remedial measures and construction works should be to minimize hazards to workers who come onto the site after the remedial actions have been completed. Thus, simple precautions are (1) to impose sufficient clean fill so that any excavation to install minor services (water, electricity, telephones) takes place only in clean material or (2) to construct concrete conduits to carry such services. It should also be noted that construction workers already face a variety of physical and health hazards from the nature of their work and the materials that they are required to handle. Working in contaminated sites can exacerbate the situation. Cement dermatitis caused by chromate impurity in cement is common. Skin burns can occur from contact with wet concrete or lime or certain additives that are alkaline. Construction workers also frequently come into contact with a range of solvents, paints, adhesives, preservatives, and fungicides and are often subjected to the stress of working under extremes of heat and cold. All these factors are likely to increase vulnerability of construction workers to additional hazards arising from contamination.

C. Ground Engineering

In order to build, either the ground must have adequate bearing capacity or the load must be carried through to ground that does, i.e., by piling. Much contaminated ground will also be filled ground and therefore likely to have low bearing capacities. Large settlements may occur. When combustible materials are present there may be cavities where areas have been burnt out, and indeed combustion may still be in progress. Conversely, physically unstable materials may be present that can expand (e.g., some steelmaking slags). The ground will also often contain old foundations, underground tanks, and disused pipework. Engineers will often prefer to see these removed rather than left in the ground (indeed, they may be required to remove tanks under U.S. legislation). Piling through the poor ground offers one solution but the presence of contaminants may adversely affect the bearing capacity of friction piles. There is always a risk that the piles may penetrate underlying strata with low permeability thus allowing contaminants to reach an underlying aquifer.

Another approach is to seek to improve the bearing capacity of the ground by consolidation.

This may be achieved by flooding with water (obviously leading to a risk of spreading contamination), surcharging, or dynamic consolidation. As the ground is compressed, any liquid or gaseous contaminants will be squeezed out. In addition, material previously above the water table may become saturated for the first time (any changes in water table can obviously present difficulties). The use of stone columns can also provide channels for mobile contaminants and must clearly be of a material that will not be attacked by chemicals in the ground (e.g., acid attack on limestone). The consequences of settlement of the site either under self-weight or newly imposed loads need to be carefully considered. This could, for example, result in a cracked water pipe and subsequent flooding.

A foundation system that does not involve digging into the contaminated ground has obvious advantages. Thus, engineered fills using materials such as fly ash may be attractive provided they are compatible with the need to prevent upward migration of contaminants.

D. Services

Most minor services can be accommodated in the top meter (frost protection of water supply pipes may be the main criterion). While individual services can be protected against the effects of toxic and aggressive substances, this is a labor intensive activity and hence prone to flaws in workmanship. The refilled trenches (granular fill is usually required) can provide channels for migration of liquid and gaseous contaminants. Reexcavation becomes a problem; will the once clean backfill have become contaminated? It is clearly preferable, therefore, to try to arrange for all services to pass only through clean material. This may not be possible with sewerage and surface water drainage systems, but care must be exercised if they are not to become channels for movement of contaminants out of the site.

E. Attack on Construction Materials

The combinations of chemicals that may be encountered means that most building materials may be subject to attack from time to time. Only ceramic materials are generally resistant, but these are frequently used in combination with other materials, e.g., cement-based or plastics, that are more vulnerable. While the construction industry is well accustomed to dealing with acids and sulfates, it is not accustomed to dealing with combinations of these with organic compounds that are likely to attack the protective systems that would ordinarily be employed. Thus, while protection is possible, the selection of appropriate means can be difficult, and there may be little independent evidence on which to make a choice. In as much as protective measures are subject to error during installation, it is better to avoid the problem whenever possible by removing materials vulnerable to attack from the aggressive environment. Monitoring for below-ground attack is generally not possible.

F. Methane and Other Gases

No one would sensibly build on a site that is actively producing large quantities of methane or other gases (although experience has shown that there are some not very sensible people about), but, if the process of decay is well advanced and the land is located so as to be potentially valuable, then the possibility of development may become attractive. Questions then arise as to how to measure reliably the quantity of gas that is being produced, bearing in mind that it is affected by many factors, and how to design a building to prevent gas entry. Gas production rates may be affected by atmospheric pressure and the position of the water table. Apparently dormant material can become active again if disturbed during site works or if it becomes wetted (it may be forced below the table by site consolidation or the water table may rise). While systems for protecting buildings are fairly easy to conceive and implement, one must also have regard to surrounding areas (gardens, outbuildings and garages, roads, and car parking areas) and to confined spaces below ground level (e.g., culverts, sewerages).

G. Combustible Materials

Combustible materials frequently occur on old industrial and filled sites, e.g., coal and coke in fuel stocking areas, refuse, oils, tars, spent oxide on gasworks (elemental sulfur). It is necessary to avoid igniting such materials during construction (the site bonfire to dispose of surplus timber) or subsequent development (underground fires have been started by the heat from furnaces in basement boiler houses and even by a small garden bonfire). This is best done by restricting access of oxygen (but very little is required to sustain slow smouldering) and providing an adequate thickness of noncombustible material between any source of ignition and the combustible material. Compartmentalization of sites using vertical noncombustible barriers may also be desirable.

ACKNOWLEDGMENTS

Figures in this chapter are reproduced by permission of the Controller of Her Majesty's Stationery Office. Crown Copyright. Building Research Establishment.

REFERENCES

1. **Harris, M. R.,** Recognition of the problem, in *Reclaiming Contaminated Land,* Cairney, T. C., Ed., Blackie, Glasgow, 1986, 1.
2. **Holdgate, M. W.,** *A Perspective of Environmental Pollution,* Cambridge University Press, Cambridge, 1979.
3. **Beckett, M. J. and Simms, D. L.,** Assessing contaminated land: U.K. policy and practice, in *Contaminated Soil,* Assink, J. W. and van den Brink, W. J., Eds., Martinus Nijhoff, Dordrecht, 1986, 285.
4. Royal Commission on Environmental Pollution, Managing Waste: The Duty of Care, 11th Report, Her Majesty's Stationery Office, London, 1985 (Cmnd 9675).
5. **Eikelboom, R. T. and von Meijenfeldt, H.,** The soil clean-up operation in The Netherlands; further developments after five years of experience, in *Contaminated Soil,* Assink, J. W. and van den Brink, W. J., Eds., Martinus Nijhoff, Dordrecht, 1986, 255.
6. California Assembly Bill 2310, Hazardous Waste: Disposal Sites.
7. New Jersey Environmental Clean-Up and Responsibility Act.
8. **Smith, M. A., Ed.,** *Contaminated Land: Reclamation and Treatment,* Plenum Press, London, 1985.
9. **Assink, J. W. and van den Brink, W. J., Eds.,** *Contaminated Soil,* Martinus Nijhoff, Dordrecht, 1986.
10. Remedial Action at Waste Disposal Sites, Report EPA/625/6-85/006, U.S. Environmental Protection Agency, Washington, D.C., 1985.
11. Slurry Trench Construction for Pollution Migration Control, Report EPA/540/2-84-001, U.S. Environmental Protection Agency, Washington, D.C., 1984.
12. Modeling Remedial Actions at Uncontrolled Hazardous Waste Sites, Report EPA/540/2-85/001, U.S. Environmental Protection Agency, Washington, D.C., 1985.
13. Guide for Decontaminating Buildings, Structures, and Equipment at "Superfund" Sites, Report EPA/600/2-85/028, U.S. Environmental Protection Agency, Cincinnati, 1985.
14. *Proc. Annu. Conf. Management Uncontrolled Hazardous Waste Sites,* Hazardous Materials Control Research Institute, Silver Springs, Md., 1980 to 1987.
15. **Cairney, T. C., Ed.,** *Reclaiming Contaminated Land,* Blackie, Glasgow, 1986.
16. **Doubleday, G. P., Ed.,** *Reclamation of Former Iron and Steelworks Sites,* Durham County Council/Cumbria County Council, Durham, 1983.
17. Superfund Strategy, OTA-ITE-252, U.S. Congress, Office of Technology Assessment, Washington, D.C., 1985.
18. **Kufs, C., Twedell, D., Paige, S., Wetzel, R., Spooner, P., Colonna, R., and Kilpatrik, M.,** in *Proc. Conf. Management Uncontrolled Hazardous Waste Sites,* Hazardous Materials Control Research Institute, Silver Springs, Md., 1980, 30.
19. **Unites, D., Possidento, M., and Houseman, J.,** in *Proc. Conf. Management Uncontrolled Hazardous Waste Sites,* Hazardous Materials Control Research Institute, Silver Springs, Md., 1980, 25.

20. **Casteel, D., Meyer, J., Ronsom, M., White, M., and Young, L.,** in *Proc. Conf. Management Uncontrolled Hazardous Waste Sites,* Hazardous Materials Control Research Institute, Silver Springs, Md., 1980, 275.
21. **Smith, M. A.,** Contamination and the built environment, in *Proc. Semin. Policy Analysis for Housing and Planning,* Planning and Transport Research Centre (PTRC) Summer Meet., PTRC, London, 1982, 37.
22. **Bernard, H.,** Love Canal 2030 AD, in *Proc. Conf. Management Uncontrolled Hazardous Waste Sites,* Hazardous Materials Control Research Institute, Silver Springs, Md., 1980, 220.
23. **Smith, M. A.,** Available reclamation methods, in *Reclaiming Contaminated Land,* Cairney, T. C., Ed., Blackie, Glasgow, 1986, 114.
24. **Hoogendoorn, D. and Rulkens, W. H.,** Selecting the appropriate remedial alternative: a systematic approach, in *Contaminated Soil,* Assink, J. W. and van den Brink, W. J., Eds., Martinus Nijhoff, Dordrecht, 1986, 1.
25. Guidance on Feasibility Studies under CERCLA, Report EPA/540/9-85/003, U.S. Environmental Protection Agency, Washington, D.C., 1985.
26. **Heaps, K. D.,** The reclamation of a disused sewage works, *Public Health Eng.,* 10(1), 213, 1982.
27. **Lowe, G. W.,** Investigation of land at Thamesmead and assessment of remedial measures to bring contaminated sites back into beneficial use, in *Proc. Conf. Management Uncontrolled Hazardous Wastes Sites,* Hazardous Materials Control Research Institute, Silver Springs, Md., 1984, 560.
28. **Stief, K.,** Long-term effectiveness of remedial measures, in *Contaminated Land: Reclamation and Treatment,* Smith, M. A., Ed., Plenum Press, London, 1985, 13.
29. **Hazaga, D., Fields, S., and Clemons, G. P.,** Thermal treatment of solvent contaminated soils, in *Proc. Conf. Management Uncontrolled Hazardous Waste Sites,* Hazardous Materials Control Research Institute, Silver Springs, Md., 1984, 404.
30. **Rulkens, W. H., Assink, J. W., and Gemert, W. J. Th.,** On-site processing of contaminated soil, in *Contaminated Land: Reclamation and Treatment,* Smith, M. A., Ed., Plenum Press, London, 1985, 37.
31. **de Leer, E. W. D.,** Thermal methods developed in The Netherlands for the cleaning of contaminated soil, in *Contaminated Soil,* Assink, J. W. and van den Brink, W. J., Eds., Martinus Nijhoff, Dordrecht, 1986, 1.
32. **de Kreuk, J. M.,** The microbiological decontamination of excavated soil, in *Contaminated Soil,* Assink, J. W. and van den Brink, W. J., Eds., Martinus Nijhoff, Dordrecht, 1986, 669.
33. **Assink, J. W.,** Extractive methods for soil decontamination: a general survey and review of operational treatment installations, in *Contaminated Soil,* Assink, J. W. and van den Brink, W. J., Eds., Martinus Nijhoff, Dordrecht, 1986, 655.
34. *Proc. Semin. Biotechnology: New Options in Land Decontamination,* European Studies Conferences, Uppingham, U.K., 1985.
35. **Landreth, R. E.,** Physical properties and leach testing of solidified/stabilised industrial wastes (NTIS Order No. PB83-147983), U.S. Environmental Protection Agency, Cincinnati, 1983.
36. **Anon.,** Getting to grips with waste solidification, *Ends Report,* 120, 11, 1985.
37. **Brugger, J. E., Yezzi, J. J., Wilder, I., Freestone, F. J., Miller, R. A., and Pfrommer, C., Jr.,** The EPA-ORD Mobile Incineration System: Present Status, in Proc. Hazardous Materials Spill Conf., U.S. Environmental Protection Agency, Edison, N.J., 1982, 116.
38. Robertsons Research International, Llandudno, Wales.
39. **Sanning, D. E.,** In-situ treatment, in *Contaminated Land: Reclamation and Treatment,* Smith, M. A., Ed., Plenum Press, London, 1985, 91.
40. **Barry, D. L.,** Treatment options for contaminated land, Atkins Research and Development, Epsom, U.K., 1982.
41. **Fitzpatrick, V. F., Buelt, J. L., Oma, K. H., and Timmerman, C. L.,** In-situ vitrification — a potential remedial action technique for hazardous wastes, in *Proc. Conf. Management Uncontrolled Hazardous Wastes,* Hazardous Materials Control Research Institute, Silver Springs, Md., 1984, 191.
42. **Van Assche, C. and Vyttebroek, P.,** Heavy metals in soils and their neutralisation, *Agric. Wastes,* 2(4), 279, 1980.
43. **Anon.,** Measures for treating soil contamination caused by hexavalent chromium in Tokyo, Bureau of Environmental Protection, Tokyo Metropolitan Government, 1980.
44. **Linfors, L. G.,** Reclamation of site of a herbicide factory, in *Proc. Conf. Reclamation of Contaminated Land,* Society of the Chemical Industry, London, 1980, F8.
45. **Childs, K.,** Management and treatment of groundwater: an introduction; In-ground barriers and hydraulic measures; Treatment of contaminated groundwater; and Mathematical modelling of pollutant transport by groundwater at contaminated sites, in *Contaminated Land: Reclamation and Treatment,* Smith, M. A., Ed., Plenum Press, London, 1985, 141.
46. **Brown, K. W., Thomas, J. D., and Green, J.,** Permeability of compacted soils to solvent mixtures and petroleum products, in Proc. 10th Annu. Res. Symp., Land Disposal of Hazardous Wastes, Report EPA 600/9-84-007, U.S. Environmental Protection Agency, Washington, D.C., 1984, 124.

47. **Anderson, D. C., Gill, A., and Crawley, W.,** Barrier-leachate compatibility: permeability of cement/asphalt emulsions and contaminant resistant bentonite/soil mixtures to organic solvents, in *Proc. Conf. Management Uncontrolled Hazardous Waste Sites,* Hazardous Materials Control Research Institute, Silver Springs, Md., 1984, 131.

48. **Beine, P. R. A. and Geil, M.,** Physical properties of lining system under percolation of waste liquids and their investigation, in *Contaminated Soil,* Assink, J. W. and van den Brink, W. J., Eds., Martinus Nijhoff, Dordrecht, 1986, 863.

49. **Lutton, R. J., Regan, G. L., and Jones, L. W.,** Design and Construction of Covers for Solid Waste Landfills, Report EPA/600/1-79-165, U.S. Army Corps of Engineers, Waterways Experimental Station, Vicksburg, Miss., 1979.

50. **Parry, G. D. R. and Bell, R. M.,** Covering systems, in *Contaminated Land: Reclamation and Treatment,* Smith, M. A., Ed., Plenum Press, London, 1985, 113.

51. **Cairney, T. C.,** Soil cover reclamations, in *Reclaiming Contaminated Land,* Cairney, T. C., Ed., Blackie, Glasgow, 1986, 144.

52. **Bradshaw, A. D. and Chadwick, M. J.,** *The Restoration of Land,* Blackwell, London, 1980.

53. **Barry, D. L.,** Hazards from landfill gas, in *Reclaiming Contaminated Land,* Cairney, T. C., Ed., Blackie, Glasgow, 1986, 223.

54. **Parry, G. D. R. and Bell, R. M.,** The impact of landfill gas on open space and agricultural restoration schemes, in *Proc. Int. Land Conference,* Industrial Seminars Limited, Tunbridge Wells, 1983, 510.

55. **Cairney, T. C.,** Long-term monitoring of reclaimed sites, in *Reclaiming Contaminated Land,* Cairney, T. C., Ed., Blackie, Glasgow, 1986, 170.

56. **Beck, W. W., Dunn, A. L., and Emrich, G. H.,** Leachate quality improvement after top sealing, Proc. 8th Annu. Res. Symp. Land Disposal Hazardous Wastes, Schuts, D. W., Eds., U.S. Environmental Protection Agency, Washington, D.C., 1982, 464.

57. **Jones, A. K., Bell, R. M., Barker, L. J., and Bradshaw, A. D.,** Coverings for metal contaminated land, in *Proc. Conf. Management Uncontrolled Hazardous Waste Sites,* Hazardous Materials Control Research Institute, Silver Springs, Md., 1982, 183.

Chapter 6

TECHNIQUES FOR THE CREATION OF WETLAND HABITAT IN COAL SLURRY PONDS

Carol S. Thompson

TABLE OF CONTENTS

I.   Introduction 92

II.   Preexisting Slurry Basins 92
        A.   Ayrshire Mine Slurry Project 93
        B.   Results 95
        C.   Benefits 96

III.   Slurry Management Techniques 97
        A.   Cyclone Separation 98
                1.   Leahy Mine Cyclone Project 98
                2.   Results 99
        B.   Multiple Discharge Method 99
                1.   Sunspot Mine Project 100
                2.   Results 100

IV.   Hydrologic and Hydraulic Considerations 100

Acknowledgments 101

References 114

I. INTRODUCTION

Coal slurry is the by-product of the coal cleaning process. It consists of fine coal particles and other constituents such as pyrite, calcite, sand, and clay ranging in size from $+60$ to -250 mesh. Pyrite is, of course, an undesirable element due to its tendency to be converted to sulfuric acid under natural weathering conditions. For this reason slurry disposal ponds have in the past been considered pollution sources to be eliminated by covering with a nontoxic soil material. The result has been creation of federal regulations requiring the coal mine operator to cover all slurry ponds with 4 ft of suitable soil material. In reality, however, 6 to 8 ft is often required in order to operate heavy, earth-moving equipment on such soft material as slurry. This means hundreds of thousands to several million dollars may be required to reclaim a single slurry pond.

Higher reclamation costs have, in recent years, caused more attention to be given to the natural soil constituents that occur in the slurry itself. Field research and experimentation by the Cooperative Wildlife Research Laboratory (CWRL) of Southern Illinois University over the last 10 years has yielded significant findings. They have discovered that if the pyrite and/or acid conditions can be eliminated from the system, the resultant material, i.e., sand, clay, and fine coal, can physically function as a soil and is not unlike many sandy or clayey soils typical of the Midwest or other parts of the U.S. The one basic element that this soil lacks is fertility which may be supplemented with artificial fertilizer and nurse crops until such time that organics in the soil become self-sustaining through plant material growth and death cycles.[1,2]

Another important element of a slurry pond is the soil moisture availability. Most slurry ponds have a substantial portion of the acreage in shallow standing water and saturated soils typical of most inland wetlands. Therefore, with the elimination of acid conditions and addition of fertilizer and plant materials, the low wet areas of a slurry pond can be converted to wetland habitat while the dryer areas may be planted to supportive upland vegetation. All of this may be accomplished with limited soil cover resulting in a substantial cost savings to the mine operator and assuring elimination of a potential pollution source. Of equal importance is the creation of wetland/wildlife habitat. It has been estimated that between 10 and 14 million acres of wetland habitat were lost between the 1950s and 1970s.[3] Although it is suspected that wetland destruction has slowed in recent years, this continual loss has resulted in a reduction of more than 50% in the original wetlands of the U.S.[3,4]

The methods for accomplishing the conversion from disposal area to wetland are varied and dependent upon the disposal site conditions, mining conditions, coal quality, preparation procedures, and amount of preplanning used before the disposal began.[5] Primary among these is the amount of preplanning and careful implementation. The other conditions can be assessed by means of a thorough site characterization and from that an appropriate treatment regime can be developed.[6] However, in order to maximize efforts in the creation of wetland, it is best to construct a complete plan for the proposed disposal site prior to initiating discharge. Several options are then available for managing the slurry to minimize treatment requirements and maximize the amount of wetland created. The remainder of this chapter will discuss three types of management practices and actual field trials of these methods. All field data collection was conducted by CWRL.

II. PREEXISTING SLURRY BASINS

There are many slurry ponds in existence today, ranging from new to decades old, which were created by what may be called the traditional disposal method. Traditional methods involve little if any preplanning or manipulation of the slurry discharge. Typically, a site (i.e., final cut impoundment, incline, or other appropriate depression) is selected to receive the slurry. A discharge pipe is then brought to the site and filling of the pond proceeds. Although the pipe may occasionally be moved a short distance to facilitate even filling or proper flow, essentially the pond is filled from a single discharge point. Due to the natural gravitational separation of the slurry constituents, a zonation effect is achieved. The larger and heavier particles drop out near the point of discharge resulting in a large mound referred to as the discharge cone. Due to the specific gravity of pyrite, the discharge cone usually contains the largest proportion of pyrite and, therefore, the greatest potential for acid production. The next lighter material is deposited in roughly the center of the pond in an area commonly referred to as the intermediate zone. The low and impounded zones occur at the far (decant) end of the pond and contain the lightest soil material including the greatest amount of clay particles and a larger amount of carbonates relative to pyrite. A detailed discussion of the zonation effect can be found in Reference 2.

A. Ayrshire Mine Slurry Project

An example of a traditional type of single discharge disposal site exists at AMAX Coal Company's Aryshire Mine near Evansville, Ind. The site was used for slurry disposal from 1980 to 1981. A detailed site characterization was conducted by CWRL in 1982 which determined that at least a portion of the site could be converted to wetland/wildlife habitat. The pond was typical of a single discharge with a discharge cone (10 acres), intermediate zone (6 acres), low zone (4.4 acres), and an impounded zone (37 acres) readily apparent.

Chemical characterization showed the discharge cone to contain vast quantities of pyritic sulfur. Values ranged from 3 to 13%. Acid/base balance showed deficit neutralization potential exceeding 100 tons $CaCO_3$eq./1000 tons over much of the zone. This translates to excessive use of lime to effectively neutralize acid generation from exposed pyrite. Such treatment levels are not uncommon for the discharge cone in a single discharge management scheme. Because of the high acid potential of the cone, which is accentuated by its porosity and low water table, and the mechanical difficulties of operating lime application equipment on this type of material, it is normally best to consider covering this zone with the usual 4 ft of soil; this was the procedure followed on the Ayrshire pond. In order to neutralize any existing acidity and to create a buffer zone between the soil and the slurry, a lime application of 40 tons/acre was applied before soil covering. Soil covering was completed during the winter of 1983, and the area was disced and seeded to a cover crop in the spring of 1984.

Early characterization of the intermediate zone showed it to contain significantly less pyritic sulfur (2 to 6%) than the discharge cone and, thus, it was manageable without soil cover. Application of approximately 15 tons of lime and sufficient fertilizer determined by standard agricultural fertility testing to allow for the growth of an initial cover crop were applied in the fall of 1983. The area was then seeded with a mixture of grasses and legumes (Table 1). The seed, lime, and fertilizer were disced in using a small tractor specifically designed to operate on slurry.

The low zone, which demonstrated an even further reduction in pyrite (1.5 to 4%), was treated in a similar manner with the exception that lime application was reduced to 10 tons/ acre. A portion of the impounded zone not yet inundated received comparable treatment, although, considerable amounts of millet were also seeded there to promote the development of good waterfowl habitat.

Both the intermediate and low zones received additional fertilizer in the spring and the fall of 1984. Furthermore, lime was administered in the fall of 1984, bringing the pond up to the anticipated level of needed neutralization potential. Both zones received approximately 30 tons/acre. The discharge cone, being covered with 4 ft of soil, needed no further treatment.

Hand planting of trees and wetland plants began in the spring of 1984. Some replanting was also done in the spring of 1985 due to losses from drought conditions during the summer of 1984 and predation on wetland plants by grazing geese. Table 2 shows the species and

Table 1
TYPICAL SEED MIXTURE APPLIED TO IMPOUNDED, INTERMEDIATE, AND LOW ZONES OF THE AYRSHIRE SLURRY WETLAND/WILDLIFE AREA

Species	Seeding rate (lb/acre)
Intermediate and low zone (10 acres)	
Winter rye (Secale cereale)	25
Foxtail millet (Seraria italica)	10
Sorgum-sudangrass (Sorgum bicolorx sudanese)	10
Birdsfoot trefoil (Lotus corniculatus)	12
Reed canarygrass (Phalaris arundinacea)	10
Redtop (Agrostis gigantia)	5
Smooth brome (Bromus inermis)	15
Orchardgrass (Dactylis glomerata)	12
Garrison creeping foxtail (Alopecurus arundinaceus)	10
Total	109
Impounded zone[a]	
Winter rye	30
Foxtail millet	10
Japanese millet (Echinochola crus-galli)	15
Sorghum-sudangrass	15
Total	70

a   Prior to adjustment of water level, 15 acres of future impounded area were exposed.

Table 2
SPECIES AND LOCATION OF TREE PLANTINGS IN AYRSHIRE SLURRY WETLAND/WILDLIFE AREA

Tree species	Zone
Bald cypress (Taxodiu distichum)	L, Imp
Green ash (Fraxinus pennsylvanica)	D, I, L, P
Sycamore (Platanus occidentalis)	I, L, P
Pin oak (Quercus palustris)	D, I, L
Red oak (Q. rubra)	D, I
White oak (Q. alba)	D, I
Tulip poplar (Liriodendron tulipifera)	D, I
Black walnut (Juglans nigra)	D, I
Loblolly pine (Pinus taeda)	D, I
Red pine (P. resinosa)	D, I
Black cherry (Prunus serotina)	D, I
Red gum (Nyssa sylvatica)	D, I, L, P
Hackberry (Celtis spp.)	D, I, L, P
Hickory (Cara spp.)	D, I
Washington hawthorn (Crataegus phaenopyrum)	D, I, P
Silver maple (Acre sacharinum)	D, I, L, P
Sugar maple (A. saccharum)	D, I
Grey dogwood (Cornus racemosa)	D
Silky dogwood (C. obliqua)	D, I, L, P
Black locust (Robinia pseudeacacia)	D, I, P

Note: D, discharge cone; I, intermediate zone; L, low zone; Imp, impounded zone; P, pond perimeter.

Table 3
SPECIES AND PROPAGULE TYPE OF WETLAND VEGETATION PLANTED IN THE AYRSHIRE WETLAND/WILDLIFE AREA

Species	Propagule type
Sago pondweed (*Potamogeton pectinatus*)	Tubers
Curley-leaved pondweed (*P. crispus*)	Plants
	Rhizomes
Common pondweed (*P. nodosus*)	Plants
	Rhizomes
Water willow (*Justicia americana*)	Plants
	Rhizomes
Yellow pond lily (*Nuphar advena*)	Tubers
Lotus (*Nelumbo lutea*)	Rhizomes
	Seeds
	Seedlings
Pickeralweed (*Pontederia cordata*)	Rhizomes
Arrowhead (*Sagittaria latifolia*)	Tubers
Burreed (*Spargamium eurycarpum*)	Rhizomes
Hardstem bulrush (*Scirpus acutus*)	Rhizomes
Woolgrass (*S. cyperinus*)	Plants
River bulrush (*S. fluviatilis*)	Rhizomes
Threesquare (*S. americanus*)	Rhizomes
Sweet flag (*Acorus calamus*)	Plants
Square spikerush (*Eleocharis quadrangulata*)	Plants
Prairie cordgrass (*Spartina pectinata*)	Rhizomes
Sedge (*Carex* spp.)	Plants
Lizards tail (*Saururus cernuus*)	Seedlings
Horsetail (*Equisetum hyemale*)	Plants
Buttonbush (*Cephanthus occidentalis*)	Plants
Japanese millet (*Echinochola crusgalli*)	Seed
Large-seeded smartweed (*Polygonum pennsylvanicum*)	Seed

location of trees planted, and Table 3 lists the moist soil vegetation and type of propagules planted in the low and impounded zones. Approximately 12,500 trees and 46,000 wetland plants were hand planted during the two seasons. Tree seedlings were obtained primarily from the state tree nursery. Some moist soil plants were purchased from commercial nurseries, but most were hand collected from the wild and other slurry projects. Local collection, although more labor intensive, has been shown to yield better survival.[1]

After initial planting, water level in the low zone was manipulated to encourage wetland plant development. The standard wetland management practice of raising water levels in the fall and lowering them in spring to allow for germination, growth, and development of the maximum number of species was accomplished by raising and lowering the spillway.[7-9] This process is very effective in increasing species diversity.

B. Results

The progress of the Ayrshire slurry pond has been closely monitored over the past 4 years in order to evaluate the effectiveness of the methods applied to this pond. The main parameters of concern are soil and groundwater acidity conditions following treatment. If these elements are within reasonable limits, past experience with other ponds has shown that plant materials will do well with little additional assistance.[10]

The results of soil pH monitoring are shown in Figures 1, 2, and 3.* In 1982, immediately

---

* Figures for this chapter appear at the end of the text.

following cessation of pumping of slurry to the pond, all zones showed relatively good pH values: range 5.0 to 6.0. However, by 1983 the top 6 in. of soil had shown a significant drop in pH (2.0 to 3.0 range) due to weathering and subsequent acidification prior to lime application (Figure 1). Soil at the 18- and 36-in. depths and in the impounded zone showed little or no effect due to isolation from the oxidation process (Figures 2 and 3).

Following lime application to the pond in 1983 and 1984, soil pH values rose steadily so that by the end of 1985 surface pHs were in the 6.0 to 7.0 range (Figure 1). Again little change was observed at lower depth.

A detailed look at the response of each exposed zone is shown in Figures 4, 5, and 6 in which the acid base accounting is represented. The intermediate zone (Figure 4) is a textbook example of the expected result of lime application. Before treatment, the potential acidity present in the slurry, primarily as pyritic sulfur, was not outweighed by the neutralization potential present in the calcium carbonates contained in the slurry. Consequently, a negative net neutralization resulted. Oxidation occurred and low pHs developed. After the addition of lime in 1983, the neutralization capacity outstripped the potential for acid production and pHs began to rise. Of course, as acid is formed and neutralized, potential acidity continues to fall, eventually approaching zero. This is the ideal result of a treatment system. The results in the low and impounded zones (Figures 5 and 6), although not as swift due to slower reactions in the saturated soils, was none the less the same.

Water quality samples collected from lysimeters showed results similar to the soils. As shown in Figures 7 and 8, groundwater pH in the top 18 in. dropped in 1983, but was followed by a gradual increase following treatment. Little change in pH was seen at the 36-in. level which remained above 7.0 throughout the study (Figure 9). This is again due primarily to constant saturation at this depth.

Rapid revegetation followed the success in treating potentially acid soil conditions. Following the first application of seed, fertilizer, and hand plantings (2 years), the discharge cone and intermediate zone showed a 70 to 100% vegetative cover. The low, seasonally inundated zone showed a 50 to 70% cover. The interspersion of wetland plant material and open water as opposed to 100% cover in this zone was planned in order to create conditions more favorable to waterfowl usage.[11] Figures 10 and 11 show aerial views of the pond from the same location in 1982 before treatment began and in 1985 following treatment. A second set of photos taken at ground level are shown in Figures 12 and 13. After two seasons of growth, the area is hardly recognizable as the same location or a slurry pond.

C. Benefits

The success of this slurry reclamation project is reflected in its benefit to the wildlife of the area. Field observations of wildlife using the pond have shown that representatives of the entire food chain have already established themselves on the pond. Top predators such as coyote, red fox, and marsh hawk use the site regularly for feeding. Shortgrass areas provide habitat for mice and voles while the tallgrass provides bedding areas and protection for white tail deer and rabbits. The wetland area is used by a variety of shore birds, such as egrets, herons, gulls, sanderlings, rails, sandpipers, killdeers, dowitchers, and yellowlegs, as well as ducks and geese, including giant Canada geese, mallards, shovelers, blue-winged teal, and coots. Table 4 gives a complete listing of wildlife observed. The excellent water quality in the impounded zone yields a constant supply of minnows, fish fry, and invertebrates for shorebird feeding, and plantings such as millet, smartweed, rushes, and sedges provide food sources for ducks and geese.

The benefits of this type of reclamation alternative can also be measured in monetary returns to the mine operator and, thus, to the consumer of electric power through reclamation cost savings. Reclamation of the Ayrshire pond to wetland/wildlife habitat, even with the

Table 4
BIOTA OBSERVED ON THE AYRSHIRE WETLAND/WILDLIFE AREA

Animals

Voles	*Microtus* spp.		
Eastern cottontail	*Sylvilagus floridanus*		
Coyote	*Canis latrans*		
Raccoon	*Procyon lotor*		
Muskrat	*Ondatra zibethica*		
Whitetailed deer	*Odocoileus virginianus*		
Mice	*Peromyscus* spp.		
Red fox	*Vulpes fulva*		
Toad	*Bufo americana*		
Leopard frog	*Rana* spp.		
Painted turtle	*Chrysemys picta*		
Crayfish	Decopoda		

Fish

Largemouth bass	*Micropterus salmoides*
Bluegill	*Lopomis macrochirus*
Carp	*Cyprinus carpio*
Blackstripe topminnow	*Fundulus notatus*

Birds — aquatic

Sandpiper	*Erolia* spp.
Common egret	*Casmerodius albus*
Great blue heron	*Ardea herodias*
Little blue heron	*Florida caerulea*
Greenback heron	*Butorides virescens*
Dowitcher	*Limnodromus* spp.
Snipe	*Capella gallinago*
Killdeer	*Charadrius vociferus*
American bittern	*Botaurus lentiginosus*
Sora	*Porzana carolina*
Blue-winged teal	*Anas discors*
Mallard	*A. platyrhyorchos*
Lesser scaup	*Aythya affinis*
Shoveler	*Spatula clypeata*
Coots	*Fulica americana*
Giant Canada goose	*Branta canadensis*
Ringbilled gull	*Larus delewarensis*

Birds — other

Morning dove	*Zenaidura macroura*
Rock dove	*Columba livia*
Barn swallow	*Hirundo rustica*
Indigo bunting	*Passerina cyanea*
Sparrow	Fringillidae
Redwinged blackbird	*Agelaius phoeniceus*
Eastern meadowlark	*Sturnella magna*
Short-eared owl	*Asio flammeus*
Marsh hawk	*Circus cyaneus*

intensive monitoring, resulted in a cost savings of approximately 70% over the traditional 4 ft of soil cover over the entire pond. Thus, the benefits to wildlife by recreating one of the most rapidly declining habitats in the country and the monetary incentives to both the operator and consumer combine to form a great reclamation tool.

III. SLURRY MANAGEMENT TECHNIQUES

An alternative to dealing with abandoned slurry ponds is to actively participate in the disposal process. By managing the placement and/or separation of desirable fractions of the slurry, further cost savings and even more wetland habitat can be created. As with most endeavors, a well-planned and -executed activity generally yields better results than one left to chance. So it is in reclaiming a slurry pond to wetland/wildlife habitat. In active mining operations, the opportunity exists to plan for the most efficient use of the fine textured silt materials within the slurry. The concept is to end up with the greatest portion of the pond possible covered with fines in order to minimize or eliminate soil cover and to minimize lime treatment.[12]

Two methods for achieving this goal include mechanical manipulation of the slurry to separate the fines from the less desirable, coarse textured materials or managing the discharge point to take maximum advantage of the natural, gravitational separation of the materials.

Although these methods have had limited field testing, the theory is quite sound and those field trials that have been conducted indicate that they can be quite successful. Of course, as with any of the discussions presented here, the methods used must be tailored to the specific project and site conditions.

A. Cyclone Separation

The use of cyclones for the separation of heterogeneous materials by means of centrifugal force is a common practice in many industries. A cyclone operates by forcing a fluid mixture such as slurry into a cylindrical, cone-shaped chamber. The material swirls against the sides of the cone, and in the process, the fine materials are separated from the coarse.[13] The fines or overflow exit at the small end of the cone and the coarse material or underflow is routed out the bottom of the assembly. The size of the cone orifice and the type of cyclone determine the specificity of separation. The classifying cyclone separates only on the basis of particle size and is, therefore, less desirable in terms of separating calcareous material from pyritic material in the slurry. On the other hand, hydrocyclones separate on the basis of specific gravity and can be adjusted to meet separation needs more precisely. If, for example, slurry analysis indicates that the calcareous materials and pyrite within the sample are of similar particle size, the classifying cyclone tends to retain both constituents in one fraction (slimes or coarse), neither of which would be ideal as an alternative soil. The classifier can be of assistance when calcareous materials are fine grained and pyrite is coarse. If a hydrocyclone is used in this same example, the desirable carbonates could be easily separated from the pyrite due to specific gravity differences between the two minerals.

Use of a hydrocyclone to separate the fine from the coarse fraction has some very distinct advantages. It is mobile and can be designed to meet the specific requirements of a given situation. Another attractive feature is that it is a piece of equipment that is normally used in the preparation of coal, thus, the pitfalls of unfamiliar equipment can be avoided. A disadvantage is that achieving the best separation requires an orifice size of 8 to 10 in. This often presents field operational/maintenance problems. The best solution to this problem may be a compromise between best possible separation and ease of field application.

Figure 14 (a to d) shows a typical sequence of slurry disposal for which a cyclone system would be ideally suited. In this example, several ponds are capped with fines from the overflow of a cyclone. While a new pond is receiving underflow material, an older pond, which is within 1 to 2 ft of hydrologically desirable capacity, is receiving the overflow as a final cap. The cap or fines are then suitable for plant growth following amendments of fertilizer and possibly lime. The overall lime requirements would be greatly reduced or eliminated depending on the quality of cyclone separation and the cap material produced. No soil covering of a high acid-producing discharge cone would be necessary. The underflow materials that have the greatest potential for acid production are then sealed from the normal weathering process by the cap and in lower areas by water and cap material.

Use of this "leap frog" approach to reclaim several ponds is particularly beneficial to active mining operations. Many mines operate over a 10 to 20 year period and subsequently must use several slurry ponds to accommodate waste disposal. Dispersal of several ponds reclaimed to a wetland/wildlife habitat around the mine area is also beneficial in diversifying the habitat of the entire area, especially in the Midwest where the post mine topography dictates a predominantly agricultural use.

1. Leahy Mine Cyclone Project

Leahy Mine in southern Illinois represents the first field application of the cyclone process for slurry reclamation. The mine was owned and operated by AMAX until it was sold to Arch Mineral Corporation in March, 1986. An existing pond of approximately 80 acres received overflow from two 24-in. classifying cyclones for a period of approximately 1 year (fall 1982 to fall 1983).

Revegetation of the pond was handled in much the same way as the Ayrshire pond. Grasses and legumes predominate in the discharge and intermediate zones, while the saturated soils and seasonally inundated zones were planted to wetland plants and grasses. In the spring of 1983, 10 acres of wet soils were planted with 19,150 wetland plants representing 12 species. Many of the species are listed in Table 3, plus alkali bulrush (*Scirpus paludosus*), slender spikerush (*Eleocharis quadrangulata*), and scouring rush (*Juncus effusus*) were used. These plantings were preceded by seeding of 8.8 acres of the moist areas with Japanese millet in the fall of 1982. Dryer sites were seeded with sudax (*Sorgham sudanese*) and winter rye (*Secale cereale*). Fertilizer was also applied at this time. In the springs of 1983 to 1985, the pond was aerially seeded and fertilized to promote a heavy permanent cover in the upland areas. In addition to the grasses and legumes shown in Table 1, red clover (*Irifolium pratense*), switchgrass (*Panicum virgatum*), reed canarygrass (*Phalaris arundinacea*), Korean Lespedeza (*Lespedeza stiulacea*), and white sweet clover (*Melitotus alba*) were included in the seed mix at various times. Fertility was determined by site testing and application rates adjusted accordingly.

2. Results

Samples taken from the surface of the pond in 1984 following capping with fines showed an average pH of 7.0 and an average net neutralization potential of -11.7 tons $CaCO_3$eq./ 1000 tons. These figures include the area of the discharge cone and represent actual field conditions without any lime applications. Although it is anticipated that some lime amendment will be necessary in the future to maintain a proper acid/base balance, it will be much less than expected on unmanaged slurry and soil cover will be unnecessary.

With a net neutralization potential of 32.5 tons $CaCO_3$eq./1000 tons, the coarse fraction of the discharge from the cyclone indicates that much of the neutralization potential within the slurry was discarded. As indicated above, this is a predictable result of using a classifying cyclone. However, even with these somewhat crude separation methods, a major reduction in needed treatment was achieved as well as a major cost reduction. Overall cost was about 10 to 15% of that for the traditional 4 ft of cover.

Figures 15 and 16 show aerial views of the pond before and after revegetation. Approximately 30 acres of moist soil vegetation is now established and serves as a nursery for wetland plants to be used on other slurry projects. It is particularly important that some propagation areas of this type exist not only for the collection of propagules but also for seed collection. Seeding of wetland plants is far less labor intensive and, therefore, less costly, but it also insures better survival because plants do not have to undergo the rigors of transplantation. It should also be noted that natural invaders such as reedgrass (*Phragmites australis*) and cattail (*Typha latifolia*) dominate portions of the site.

The advantages of cyclone separation are clearly evident from this field trial. However, much work remains to be done in refining the technique and incorporating the use of a hydrocyclone into an operating mining situation. Although the leap frog approach was not employed at the Leahy Mine due to mining operations constraints, it could easily have applied to the long-term plans for this facility and does represent a viable alternative for many mining operations. The disadvantage is that a slurry disposal plan of this type, by necessity, involves accurate, long-term planning and must incorporate the expertise of coal preparation and planning engineers, hydrologist, biologist, and agronomist in a team effort to assure success.

B. Multiple Discharge Method

The natural tendency for slurry constituents to segregate into various fractions by gravitational separation can also be used to maximum advantage. Again, the goal is to retain as much good soil material (i.e., fine-textured silts) as possible over the largest area of the surface of the pond. This may be accomplished by employing a variation of the traditional discharge method. If the discharge pipe is not left in one position, but is moved around the perimeter of the pond throughout the life of the discharge, then fines can be expected to accumulate over the surface of the majority of the pond.[12] Figure 17 (a to f) shows a schematic of such an operation. After a few successive moves of the pipe, the fines begin to cover less desirable coarse materials. Each consecutive move of the pipe, thereafter, tends to cover a previously created discharge cone with silt/clay fines. By the time the end of the useful life of the pond has been reached, the majority of the acid-producing material has been covered with a suitable growth media. The multiple discharge method results in one small discharge cone (the final discharge point) and possibly a small fringe around the extreme outer edge of the pond that would require soil cover. The outcome is, then, less soil cover, smaller lime requirements, and more productive wetland habitat.

1. Sunspot Mine Project

In 1982 a slurry project employing the multiple discharge method was begun at AMAX Coal Company's Sun Spot Mine in west central Illinois. The area to receive discharge was an old, existing slurry pond. Following a complete hydrologic investigation by CWRL, the capacity of this pond was increased by means of a levee to accommodate the final slurry output from the mine. The proposed final wetland/wildlife area was approximately 80 acres. Discharge was begun in the fall of 1982. Due to the early shutdown of the mine, discharge was discontinued in the fall of 1983 rather than late 1984 as previously planned. However, during the ensuing year the slurry was managed on schedule and as planned. The discharge pipe was moved around the perimeter of the pond 11 times resulting in approximately one half the margin being traversed.

2. Results

Although the project was not taken to completion, the intended results were well on the way to fruition when the project was halted. The discharge management resulted in the creation of slime flats downslope from the discharge zone that correspond to the low and permanently impounded zones of the Ayrshire pond. The impounded zone was characterized by positive net neutralization potentials, indicating no lime amendment was necessary, and low zone test results showed the need for a minimal lime application of 5 to 12 tons/acre. Depth of slime deposits varied from 8 in. to 6 ft. Due to the fact that the second half of the project was not completed, the discharge cones did not receive the benefit of slime covering deposits (see Figure 17 and accompanying text). Much of the area scheduled for slime cover remained exposed and often showed net neutralization potentials in excess of 100 tons $CaCO_3$eq./1000 tons. Therefore, approximately 40 acres of acid-producing discharge cone had to be covered.

Overall, however, this project was considered successful in demonstrating the potential for the multiple discharge method. Slime flats were created over one side and central portion of the site as predicted by this method. Although the remainder of the process is still to be tested, this field trial indicates the final portion of the test would have yielded favorable results. This project did produce 40 acres of seasonally inundated, moist soil land suitable for wetland vegetation. The wetland will receive approximately 12,000 hydrophyte propagules in addition to hand seeding of some sedges, rushes, and grasses in the spring of 1986. The 40-acre upland zone around the perimeter of the pond will be planted to grasses and trees.

IV. HYDROLOGIC AND HYDRAULIC CONSIDERATIONS

Proper hydrologic design is as important to the success of a slurry wetland project as correct placement of materials in the pond. An adequate water supply is needed to sustain and promote the growth of diverse wetland vegetation. This means saturated soils, allowing for some seasonal variation for dryer and wetter conditions, must be assured on a permanent basis. This is not possible in all slurry pond locations or may require some additional mechanical or physical control to insure it.

Whether a proposed project is on an existing pond or on some future disposal site, the first order of business before embarking on reclamation is to assess the hydraulics of the area. This should include a watershed analysis, i.e., how much water can be expected to come into the pond on an annual basis from the surrounding area, annual rainfall, the groundwater elevation in the area, and any contributions made by springs or seeps. From these facts, a determination of the likelihood of sustaining a wetland can be made. Or, it may be determined that additional water from a surrounding area should be channeled in or that raising the dam on an existing impoundment might provide needed water.

When preplanning for a future project, hydrologic considerations are of upmost importance. The maximum elevation of slurry within the pond in order to maintain high groundwater elevation must be calculated and vigorously adhered to during actual slurry disposal. There is always a temptation to acquire a few more inches or one more foot of disposal space. This can spell disaster for a wetland project and should be avoided. However, a hydraulic analysis may show that, in order to maintain a wetland, so much slurry capacity may be given up, that seeking additional disposal room elsewhere may not be economically prudent or physically possible. In this case, a wetland may be out of the question. In any case, the hydraulic design can determine up front the feasibility of a project.

The degree of long-term water control on a pond may also be determined by the operator. In order to produce the greatest diversity of wetland hydrophytes, an annual drawdown capability is desirable.[14] This may be accomplished by placing a control structure at the outlet of the pond. It can then be operated manually to control the water level in the pond at will. Although long-term water control may not be desirable for the mine operator, it is advisable to have this control during the revegetation phase in order to promote rapid plant establishment and diversity from the start. After vegetation establishment, natural fluctuations will maintain the vegetation, but not necessarily at the peak diversity seen in a managed situation.

ACKNOWLEDGMENTS

AMAX Coal Company and Restoration Resources gratefully acknowledges the pioneering work of the Cooperative Wildlife Research Laboratory of Southern Illinois University in developing the wetland development techniques used in the projects described in this chapter. The research efforts of Willard Klimstra, Jack Nawrot, Scott Yaich, David Warbuton, and others at CWRL over the past decade have given the coal industry another valuable tool in the reclamation process.

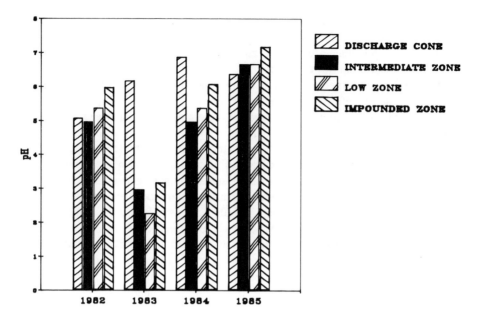

FIGURE 1.   Ayrshire slurry wetland average soil pH — 6 in.

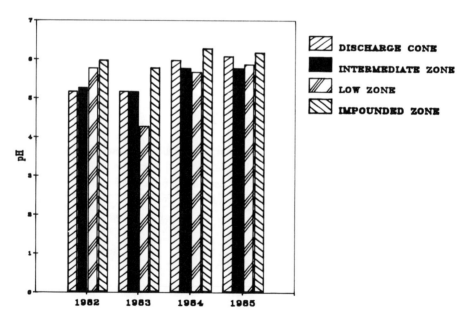

FIGURE 2.   Ayrshire slurry wetland average soil pH — 18 in.

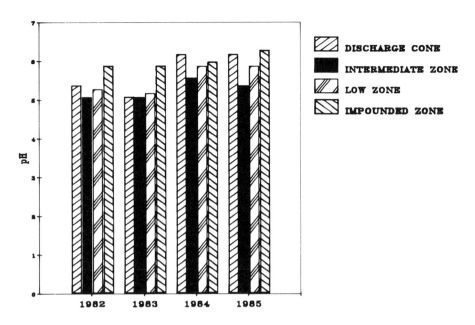

FIGURE 3.   Ayrshire slurry wetland average soil pH — 36 in.

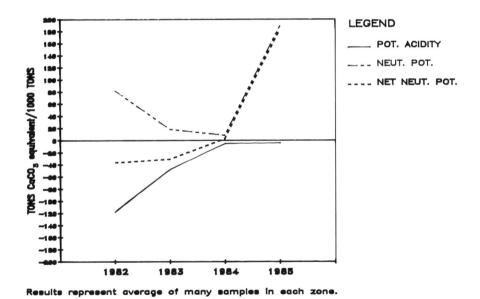

FIGURE 4.   Ayrshire slurry wetland acid/base balance at 6 in. — intermediate zone.

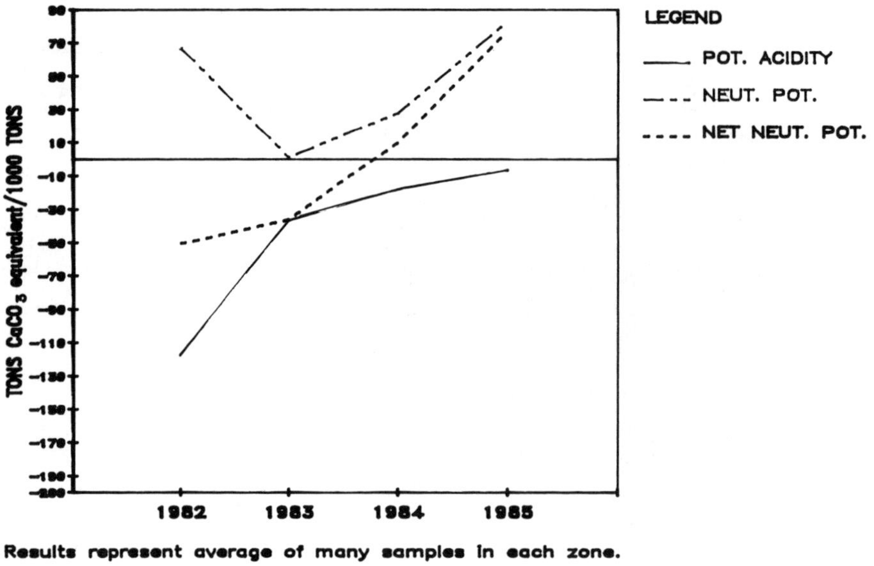

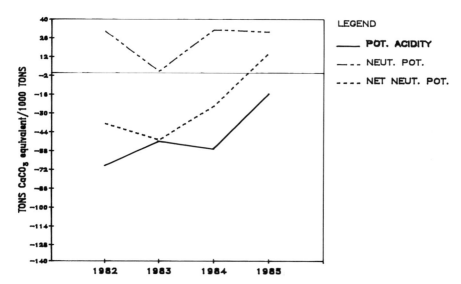

FIGURE 5.   Ayrshire slurry wetland acid/base balance at 6 in. — low zone.

FIGURE 6.   Ayrshire slurry wetland acid/base balance at 6 in. — impounded zone.

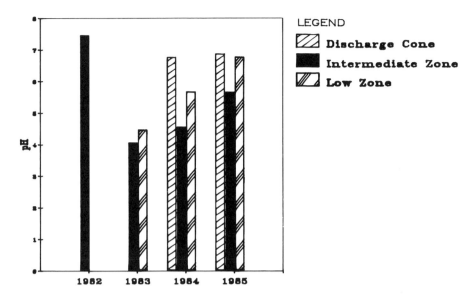

FIGURE 7.   Ayrshire slurry wetland — average groundwater pH at 6-in. depth, 1982 to 1985.

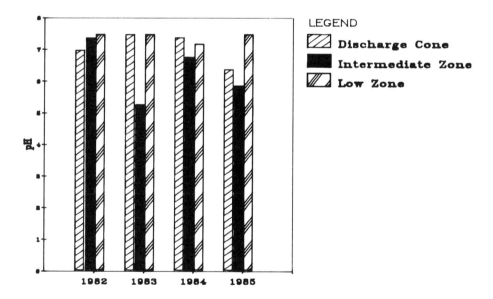

FIGURE 8.   Ayrshire slurry wetland — average groundwater pH at 18-in. depth, 1982 to 1985.

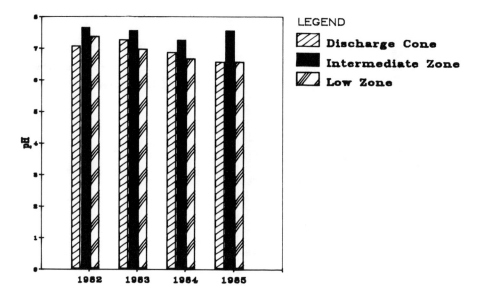

FIGURE 9. Ayrshire slurry wetland — average groundwater pH at 36-in. depth, 1982 to 1985.

FIGURE 10. Aerial view of Ayrshire Slurry Pond in 1982 prior to treatment and revegetation.

FIGURE 11. Aerial view of Ayrshire Slurry Pond in 1985 following treatment and revegetation.

FIGURE 12. Ground level view of low zone on Ayrshire Slurry Pond in 1982 prior to treatment and revegetation.

FIGURE 13. Ground level view of low zone on Ayrshire Slurry Pond in 1984 following treatment and revegetation. Note trees in background for orientation with Figure 12.

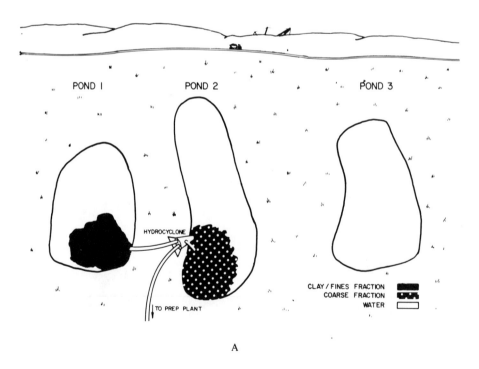

FIGURE 14. Schematic of cyclone separation and use of fines as capping material to create ideal wetland soils.

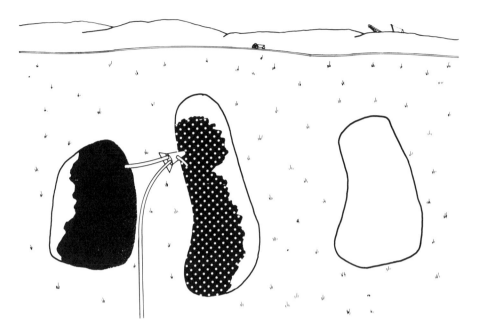

FIGURE 14B

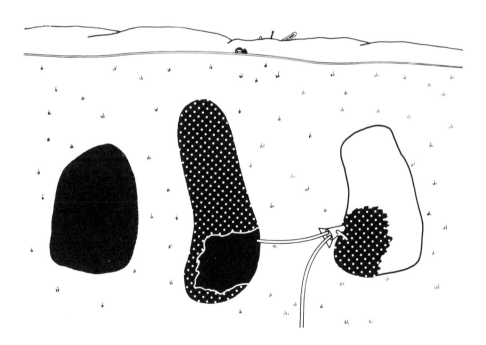

FIGURE 14C

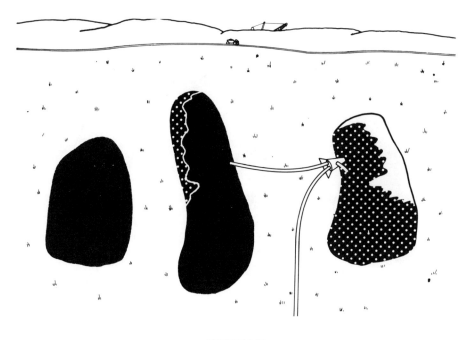

FIGURE 14D

FIGURE 15.   Leahy Mine Slurry Pond in 1982 prior to treatment and revegetation.

FIGURE 16. Leahy Mine Slurry Pond in 1985 following treatment and revegetation. Note location of coal storage silos and road to right of photo for comparison with Figure 15.

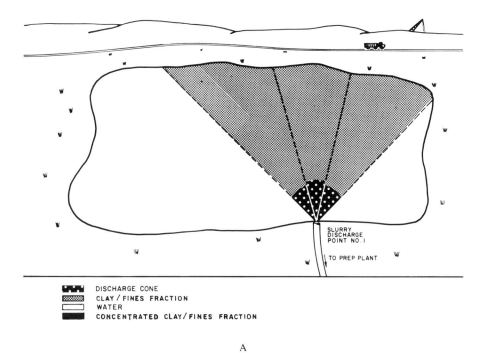

A

FIGURE 17. Schematic of the multiple discharge method of slurry management.

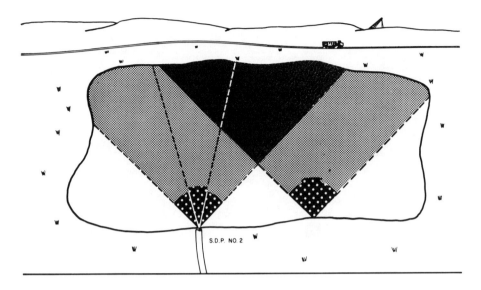

FIGURE 17B

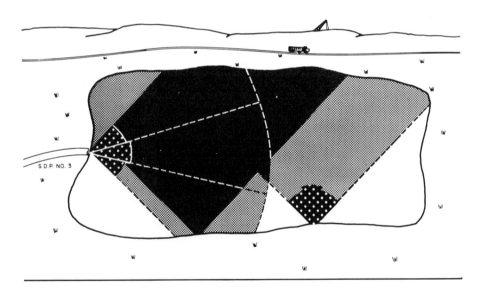

FIGURE 17C

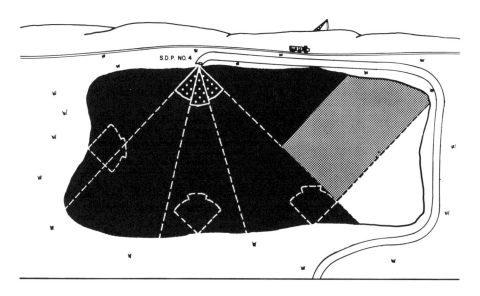

FIGURE 17D

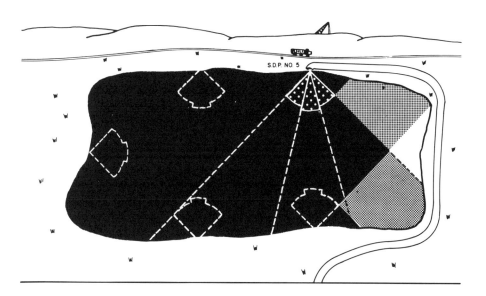

FIGURE 17E

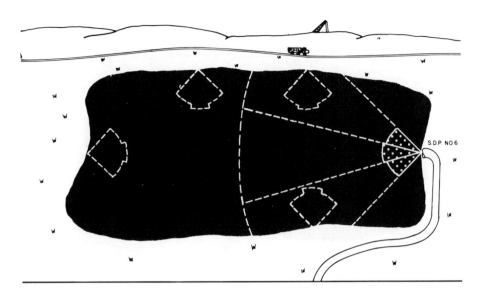

FIGURE 17F

REFERENCES

1. **Warburton, D. B., Klimstra, W., and Nawrot, J.,** Aquatic macrophyte propagation and planting practices for wetland development, in *Wetland and Water Management on Mined Lands,* Brooks, R. P., Samuel, D., and Hill, J., Eds., The Pennsylvania State University, University Park, Pa., 1985, 139.
2. **Nawrot, J. R.,** Wetland development on coal mine slurry impoundments: principles, planning, and practices, in *Wetlands and Water Management on Mined Lands,* Brooks, R. P., Samuel, D., and Hill, J., Eds., The Pennsylvania State University, University Park, Pa., 1985, 161.
3. **Frayer, W. E., Monahan, T. J., Bowden, D. C., and Graybill, F. A.,** Status and trends of wetlands and deepwater habitats in the contiguous United States, 1950's to 1970's, Department of Forest and Wood Sciences, Colorado State University, Fort Collins, 1983.
4. **Heimlich, R. E. and Langner, L. L.,** Swampbusting in perspective, *J. Soil Water Conserv.,* 41(4), 219, 1986.
5. **Nawrot, J. R. and Yaich, S.,** Slurry discharge management for wetland soils development, in *Proc. Symp. Surface Mining, Hydrology, Sedimentology and Reclamation,* Graves, D. H., Ed., University of Kentucky, Lexington, 1982, 11.
6. **Yaich, S. C., Nawrot, J., and Klimstra, W.,** Slurry impoundments: resource information for reclamation planning, presented in Mini Course 7 at the Natl. Symp. on Surface Mining, Hydrology, Sedimentology and Reclamation, University of Kentucky, Lexington, 1984.
7. **Schemnitz, S. D.,** *Wildlife Management Techniques Manual,* The Wildlife Society, Bethesda, Md., 1980, 381.
8. **Weller, M. W.,** Wetland habitats (theme paper), in *Proc. Symp. Wetland Functions and Values: The State Of Our Understanding,* Greeson, P. E., Clark, J. R., and Clark, J. E., Eds., 1978, 210.
9. **Teskey, R. O. and Hinckley, T. M.,** Impacts of Water Level Changes on Woody Riparian and Wetland Communities, Vol. 4, U.S. Fish and Wildlife Service/Office of Biological Service — 78/87, 1978, 1.
10. **Nawrot, J. R.,** Stabilization of slurry impoundments without soil cover: factors affecting vegetation establishment, in *Proc. Symp. Surface Mining, Hydrology, Sedimentology and Reclamation,* Graves, D. H., Ed., University of Kentucky, Lexington, 1981, 469.
11. **Kaminski, R. M. and Prince, H. H.,** Dabbling duck and aquatic macroinvertebrate responses to manipulated wetland habitat, *J. Wildl. Manage.,* 45(1), 1, 1981.
12. **Nawrot, J. R., Yaich, S. C., and Klimstra, W. D.,** Enhancing reclamation through selective slurry disposal, in Proc. 92nd Annu. Meet. Illinois Mining Institute, Springfield, Ill., 1984.
13. **O'Brian, E. J.,** Water-only cyclones: their functions and performance, *Coal Age,* January, 110, 1976.
14. **New, J.,** Marsh development for wildlife, Indiana Department of Natural Resources Management Series, No. 8, 1976.

Chapter 7

EVALUATION OF STRIP PITS AND PONDS FOR PHYSICAL MANIPULATION TO INCREASE WETLANDS AND IMPROVE HABITAT IN SOUTHWESTERN INDIANA

Daniel E. Willard

TABLE OF CONTENTS

I.   Introduction 116

II.   Background 116

III.   Fisheries and Wildlife Use of Wetlands 118

IV.   Management Tactics 118

V.   Aerial Inventory Methods 120

VI.   Wetland Succession 121

VII.   Opportunity 121

References 121

I. INTRODUCTION

This paper addresses three general areas: (1) the availability of coal mining-generated water bodies, particularly wetlands, for fish and wildlife habitat, (2) the potential use of wetlands for fish and wildlife, and (3) the role of natural succession in the wetlands. Embedded in these scientific topics, I discuss the role of cumulative regional additions to the wetland acreage of Indiana as mitigation for our past large losses of wetlands. Finally, I will argue that scientists now have the expertise, interest, and financial resources to expand our development of wetlands and other aquatic resources in the mined areas of southwestern Indiana.

National and state interest now focuses on restoration of wetlands, development of new wetlands, and mitigation of wetland loss. The draft research plan of the U.S. Environmental Protection Agency[1] emphasizes these areas and adds the effects of cumulative wetland loss and gain. In 1985, the Indiana legislature authorized the expenditure of a $350,000 match to U.S. Fish and Wildlife Service (USFWS) funds to complete the National Wetlands Inventory of Southwest Indiana, an area including the mine-caused lakes discussed here.

II. BACKGROUND

Surface mining has formed over 5000 water bodies in southwest Indiana. The U.S. Geological Survey[2] (USGS) reported on the physical, and to some extent the biological, characteristics of approximately 1000 of these lakes (Figure 1). The 1000 lakes ranged from 0.5 to 344 acres in size, and all occurred on nonactive mine land. Of these, the USGS sampled 287 lakes for pH and special conductance. Approximately 80% of the lakes sampled had pH readings between 6 and 9. All 15 lakes with pHs between 2.5 to 3.9 were reported clear and free of plant and animal life. Though the USGS summary suggests that little shoreline vegetation existed, inspection of their data shows that some combination of trees, grasses, weeds, or bushes were generally present. Specific conductance ranged from 99 to 3800 μmhos/cm at 25°C; however, 70% of the measurements fell within the range of 250 to 1000 μmhos/cm. Of the USGS lakes, 64 occur in Greene-Sullivan State Forest, an area of research focus for this discussion. The Indiana Department of Natural Resources lists over 112 named lakes in the 16 square miles of Greene-Sullivan State Forest. The USGS gazetteer lists size and shoreline development (as defined by Wetzel[3]) for these lakes as well, but does not provide any information on littoral zones.

The USGS has done considerable study on the stream biota in the coal mining region of Indiana,[4] but little on the ponds. The results showed reduced benthic fauna in streams draining mine areas, particularly unreclaimed areas. The USFWS produced a survey of fish and wildlife management techniques[5] for reclaimed land. The USFWS provided considerable information on upland habitat, fish habitat structures such as nest boxes, and compiled extensive lists of suppliers of plant materials and animals. They failed to examine the construction of wetlands as part of the mine reclamation plan.

Several Indiana University graduate students have conducted their dissertations working on the limnology of coal-mine lakes in this area. Smith[6] studied the water chemistry, plankton populations, and fish populations in several lakes in Pike and Gibson Counties. He compared biological food webs of lakes of varying pH in an attempt to predict recovery processes. Unfortunately most of his sites have been remined. Currently David Baas is working on diversity in zooplankton communities. He finds that mine lakes of pH 3 to 4 have much lower diversity than lakes of similar pH in North Florida. He is attempting to determine why.[6a]

The State of Indiana has considerable work proceeding on mine-caused water bodies. The Reclamation Division of the Department of Natural Resources has a large project near

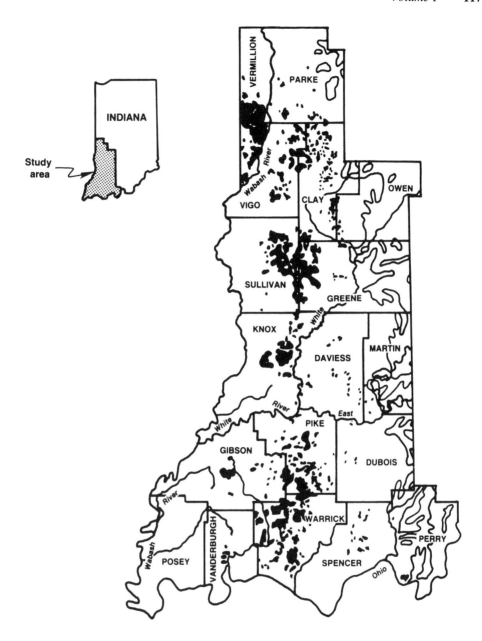

FIGURE 1. Coal field Indiana. Adapted from the U.S. Geological Survey and the Indiana Geological Survey. Mine lakes found throughout mined area. This study concentrates on the area along the Green-Sullivan County borders. Small dark areas = surface mines; large open areas = area underlain by coal bearing rocks of the Pennsylvania system; large dark areas = extensively mined area.

Jasonville, Ind. Similarly, the Indiana Department of Natural Resources has been conducting fisheries habitat studies in the Greene-Sullivan State Forest.[6b]

Knopka and colleagues[7] have studied the microbiological communities on Reservoir 29 in Greene-Sullivan. Reservoir 29 also was sampled by USGS earlier. Reservoir 29 floods abandoned gob piles and coal-washing areas. The USGS reported a pH of 3.0 in the outfall in 1979. They report that pH in the lake during spring runoff drops to 2.8, but rises to 3.5 by fall due to the sulfate reduction activities of a thin layer of bacteria on the bottom 6.8

m down. They report an improvement of about 0.1 pH units annually, and this noted the increase of frogs and corixids in the lake.

Active mining companies in Indiana have large reclamation programs in response to state and national regulations. Peabody Coal has built large lakes that have good fisheries and waterfowl populations on its Dugger Mine adjacent to Greene-Sullivan State Forest on the west. AMAX, working with Southern Illinois University, has successfully vegetated a slurry pond just north.

III. FISHERIES AND WILDLIFE USE OF WETLANDS

Indiana has lost most of its historic wetlands. The National Wetlands Inventory is not complete for Indiana, but the inventory shows[8] a loss of 99% for Iowa, an agricultural state similar to Indiana. It also lists a 71% loss for Michigan. The preamble to the proposed Indiana Wetland Protection Bill suggests that 55% of Indiana wetlands has been lost. Ongoing research comparing the original survey records to existing maps for Greene County, Ind. indicates that approximately 70% of the bottomland hardwoods has gone. This has serious implications for fisheries and wildlife in the area.

Most North American vertebrates use wetlands and many require them. A review of various field guides shows that 909 species of vertebrate animals require wetlands at some time in their lives if they are to survive (e.g., tree frogs). Similarly, 1130 species use wetlands to maintain higher population levels (e.g., white-tailed deer).

The strip pit area near Linton, Ind. primarily in the Greene-Sullivan Forest, has long been the most famous fishing area in Indiana.[9] Clearly, the area already contains some ponds of great fishery value. Similarly, beaver, deer, and waterfowl now use some of the area. No study exists about the nongame species use of the area; however, much of it is of difficult access, diverse topography, and varied water regime providing a broad range of habitats.

The fisheries and wildlife values of wetlands proved to have values sometimes far removed from those wetlands. For example, during spring and fall, almost the entire midwestern population of sandhill cranes (*Grus canadensis*), numbering about 12,000, gather at Jasper-Pulaski Wildlife Refuge in northern Indiana. They nest in Canada west to Manitoba and east into Ontario. They winter scattered along the coast of the Gulf of Mexico. In October 1982, thousands of visitors from all over North America came to observe them. Similarly, bass fishing on the available strip pit lakes has users far removed from the immediate vicinity who come to fish.

Thus, for the purpose of this discussion, the strip area contains lakes and surrounding areas of three general kinds. Some already have high value of animals. A second group may have water quality problems that make immediate reclamation unfeasible, but which may improve with time. A third category has low present populations, but may respond to economically and ecologically sound management programs.

IV. MANAGEMENT TACTICS

Generally, five physical or biological elements exist to enhance the carrying capacity of these lakes, some clearly more practical to manipulate than others. The manager should consider each. These elements are (1) water regime, (2) bottom contour, (3) cover, (4) water quality, and (5) species availability.

I will consider the first three in some detail. The last two require somewhat less discussion here. The preceding paragraphs suggest that lakes of sufficient water quality already exist and that many species already exist in the area. Others can be stocked.

Water regime includes both the temporal and elevational elements of seasonal changes in water level. Southwestern Indiana receives about 100 cm of precipitation throughout the

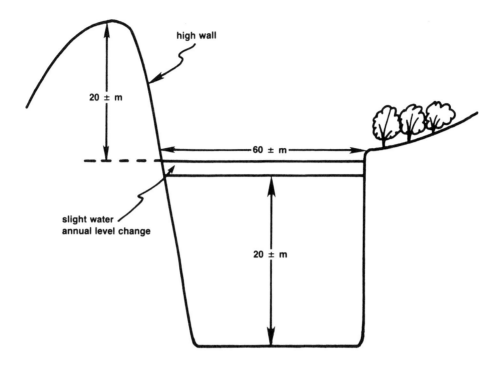

FIGURE 2. Generalized cross-section of pre-SMCRA final cut pond.

year. The heaviest rains, coupled with ice and snow melt, occur in the spring. The flood level and date of highest level vary widely between late February and May. Generally, August and September are relatively dry, but again much variability exists. No records exist for the lakes resulting from abandoned coal mines but the water regime in many lakes involves changes of a few meters or so in water level over a period of a year. Most of the lakes in the Greene-Sullivan State Forest and most in the entire mine area of Indiana have neither surface inlets or surface outlets.[2] They receive most of their water from groundwater with additional input from local surface runoff during storms. Outflow consists of groundwater and evapotranspiration. This regime is probably satisfactory for wetland habitat development and not easily modifiable.

The bottom contours throughout are generally similar. The lakes result from groundwater filling of final cuts left before the introduction of the Surface Mining Control and Reclamation Act of 1977 (SMCRA). These cuts are steep sided, approaching vertical, trenches ranging to 40 m deep in some lakes. Most final cut lakes reach to at least 12 m in depth. Some lakes, such as Reservoir 29, incorporate gob and slurry areas that are much shallower, but are limited for fisheries or wildlife use by poor water quality (Figure 2). These final cut lakes resulting from mining operations before 1977 have little or no littoral zone. Primary production must come from plankton, and nutrients are readily lost to the benthos and never recycled. The great depth of the lakes and lack of littoral zones prohibit the growth of macrophytes. Thus, the only cover occurs from terrestrial shrubs and trees that overhang the water. When hedges or shallows do occur, they tend to develop simple wetland communities dominated by emergent or submergent aquatics depending on the water depth. The emergent communities seldom exceed 20 m².

The surface mining regulations of 1977 allow active mines to retain final cut lakes, but require mine operators to grade a gentle slope of at least 4 to 1 around and into the lake. Generally, the operator simply grades the highwall into the lake, creating a littoral slope on both the terrestrial and aquatic sides of the water line. These littoral zones quickly develop

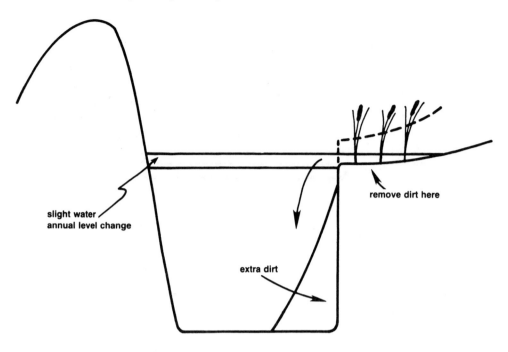

FIGURE 3.   Generalized cross-section showing simple grading necessary to increase wetland area in a pre-SMCRA final cut pond.

submerged aquatic and emergent vegetation zones. Regular observation at the Peabody Dugger Mine site near Greene-Sullivan shows abundant waterfowl and gamefish populations. In short, the limitation for habitat on many lakes could be relieved by known proved methods, simply the creation of littoral zones.

Some lakes will develop more extensive wetlands if managers simply provide additional littoral zone. This requires grading the soil surface to an elevation approximately equivalent to the summer low water level. Because of the already great depth of these ponds, the extra soil can be allowed to fall to the bottom with little habitat loss. Observation on many abandoned lakes, as well as on post-SMCRA developed lakes, shows that emergent aquatic species quickly colonize available littoral zones (Figure 3).

V. AERIAL INVENTORY METHODS

Considerable research is proceeding and data exist on coal mine lakes, but each researcher has developed his own interests and opportunities. While we know few wetlands now exist, we do not know which lakes are most suitable for wetland development. The result of the varied research activities suggests that sufficient data exist on a known set of lakes to use these lakes to calibrate or provide ground truth for an aerial survey designed to select lakes for wetland creation. Because of the high number and difficult access to many lakes, low level aerial photographs taken during late spring and early fall may be the cheapest and most efficient method of determining which lakes are candidates for enhancement through littoral zone creation.

In late spring, the water levels are usually highest. Some lakes rise up to 700 cm during highwater periods. During late summer, the water usually falls to its lowest. For many lakes with nearly vertical sides, the horizontal extent of the lake surface will differ little between these photographs taken on these two dates. Those with more gradual slopes will show not only a larger horizontal extent in spring, but may show the presence of floating and emergent aquatic macrophytes in the littoral zone. These plant communities occur in water of narrowly restricted depths and have easily recognized images.[10] With the proper film and filter combination, plankton type and productivity can be roughly estimated.

Other publicly available imagery is somewhat useful. The National Wetland Inventory photography and classification draft materials became available in 1986. The system does not clearly delineate vegetation units less than 1 acre and consequently does not provide the resolution to identify aquatic communities specifically necessary for these purposes. The delineation of 1 acre or larger wetlands will help focus our interest. The Agricultural Stabilization and Conservation Service aerial photographs are available approximately every 10 years, but only in black and white. They do help to identify temporal changes in lake morphology and vegetation.

VI. WETLAND SUCCESSION

Wetlands change over time. Study of existing ASCS photographs compared to newly taken photographs will give some indication of the rates and sorts of stages during coal mine wetland succession. Naturally occurring wetlands do not change in a deterministic way but have a wide range of potential paths depending on the fortuitous juxposition of stochastic events at particular times. Some of the pits could become important additions to the fish and wildlife habitat of Indiana. Even granted the aggressive adaptibility of the remaining species, combinations of events outside the homeostatic regulatory ability of the system are possible, if not likely. Little is known about the ecological regulatory system of coal-mined wetlands. Scientists simply have no model for wetland succession in strip pits.[11]

VII. OPPORTUNITY

The abandoned mine lands office now has sufficient funds to begin a strong program toward enhancing strip pits as habitat. The scientific knowledge and practical experience necessary have ripened. However, all pits are not created equal. Low level aerial photography can greatly assist in choosing the most ecologically and economically sound places to start.

REFERENCES

1. **Zedler, J. B. and Kentula, M. E.,** *Wetlands Research Plan,* U.S. Environmental Protection Agency, ERL Corvallis, Ore., 1985.
2. **Bodo, L. L.,** Gazetteer of Coal-Mine Lakes in Southwestern Indiana, USGS Water Resources Investigations 79-57, Indianapolis, Ind., 1979.
3. **Wetzel, R. G.,** *Limnology,* W.B. Saunders, Philadelphia, 1975.
4. **Wangsners, D. T.,** Reconnaissance of stream biota and physical and chemical water quality in areas of selected land use in the coal mining region, southwestern Indiana 1979—1980, USGS open file report 82-566, Indianapolis, Ind., 1982.
5. **Leedy, D. L.,** Coal Surface Mining Reclamation and Fish and Wildlife Relationships in the Eastern United States, Vols. 1 and 2, FWS/OBS-80/24, 1981.
6. **Smith, R.,** Acid Mine Pollution Effects on Lake Biology, Masters' thesis, School of Public and Environmental Affairs, Indiana University, Bloomington, 1973.
6a. **Baas, D.,** personal communication.
6b. **Andrews, S.,** personal communication.
7. **Knopka, A., Gyure, R., Doemel, A., and Brooks, W.,** Microbial Sulfate Reduction in Extremely Acid Lakes, Ivater Resources Research Center, Purdue University, West Lafayette, Ind., 1985.
8. **Tiner, R. W.,** Wetlands of the United States: Current Status and Recent Trends, National Wetlands Inventory, U.S. Fish and Wildlife Service, U.S. Department of the Interior, Washington, D.C., 1984.

9. **McClane, A. J.,** *Standard Fishing Encyclopedia and International Angling Guide,* Holt, Rinehart & Winston, New York, 1965.
10. **Lillesand, T. M. and Kiefer, R. W.,** *Remote Sensing and Image Interpretation,* John Wiley & Sons, New York, 1970.
11. **Willard, D. E.,** Progress report on restoration and development of wetlands on two south Chicago properties, in *Proc. National Wetland Assessment Symposium,* Kusler, J. A. and Riexinger, P., Eds., Omnipress, Madison, Wis., 1985.

Chapter 8

SALT MARSH RESTORATION: LESSONS FROM CALIFORNIA

Joy B. Zedler

TABLE OF CONTENTS

I.   Introduction 124

II.   Defining Restoration Goals 124

III.   Hydrological Planning 127

IV.   Experimentation 129

V.   Assessment of Restoration Success or Failure 130

Appendix 1 133

Appendix 2 135

References 136

I. INTRODUCTION

The salt marshes of America have all been disturbed, directly through effects of dredging and filling, or indirectly from hydrological modifications elsewhere in the watershed. Nowhere in the U.S. have these impacts been felt more rapidly and extensively than in southern California. The human population density is among the highest in the nation, with the cities of Los Angeles and San Diego in California and Tijuana in Baja California continually expanding. Since settlement about 200 years ago, the coastal region has accommodated over 20 million people. By the year 2000, the density of California is expected to increase by nearly 7 million, with most of the growth in the southern portion of the state. Because the initial area of salt marsh was small, the pressures of development have had and will continue to have an enormous impact on salt marsh habitats. In some coastal embayments, no traces of salt marsh remain; for the region as a whole, only about 25% of the coastal wetland area is left. The impact has been to endanger wetland-dependent species with extinction.

A dilemma exists for a region and a nation where development and wetlands compete for space. Federal legislation protects the habitats from dredging and filling operations (Section 404 of the Clean Water Act) and the rarest species from having their habitats impaired (Endangered Species Act). In California, State laws protect additional rare species, and the Coastal Act regulates development adjacent to coastal wetlands. However, these and other layers of legislation do not prevent further wetland destruction. Trade offs that allow further wetland losses in exchange for restoration of degraded salt marsh are still possible.

The continuing shrinkage of habitat area is justified by some as an exchange of quality for quantity. Portions of existing wetland habitat can be developed if the remaining marsh is enhanced or restored. Rarely is it acknowledged that restoration is a hope, not a guarantee. Mitigation, or "reduced impact" through restoration, thus becomes a license to develop more and more of the diminishing salt marshes of the region. What keeps alive the myth that "quality replaces quantity" is not a wealth of restoration success stories but the lack of documentation on restoration failures.

This chapter explores a broad range of guidelines that have developed from long-term research on wetlands, from observing restoration failures, and from literature review.[1-6] The restoration literature (cf. the periodicals, *Restoration and Management Notes* and *Proceedings of the Annual Conferences on Wetlands Restoration and Creation*) includes many wetland projects from the eastern U.S., while many California salt marsh examples are evaluated here for the first time (Appendix 1). Recommendations for goal definition, hydrological planning, experimentation, and assessment of success are provided. The concept of "ecotechnology" is developed as the careful manipulation of habitats to achieve desired ecosystem characteristics.

II. DEFINING RESTORATION GOALS

Intuitively, the goal of restoration should be to replace what has been lost. For several reasons, this goal is difficult to define. First, coastal salt marshes are far from being stable ecosystems. Their topography changes with sediment deposition and erosion; their species composition responds to variations in tidal and riverine influences; their rates of productivity and nutrient cycling are driven by fluxes of materials and organisms, as fresh and/or tidal water floods and drains the intertidal zone. With such dynamic structure and function, it is difficult to set a specific historic time or condition as a target for salt marsh restoration.

Second, because most systems have undergone substantial disturbance, it is often impossible to define what has been lost. Even with evidence of what type of salt marsh ecosystem existed on a site prior to its degradation, it may not be possible to recreate it because the necessary hydrology has been permanently altered. (For example, restoration of small freshwater marshes adjacent to salt marshes may be impossible where pumping from wells has lowered the water table or dried up the springs that once supported small fresh or brackish marshes.) At the same time, altered hydrology or topography at one site may provide opportunities to fulfill restoration objectives that have been specified elsewhere in the region. (For example, the potential to discharge treated wastewater near a salt marsh may make it possible to create fresh or brackish marshes where none occurred naturally.) Finally, because salt marshes are so dynamic, it may be impossible to achieve or maintain a specific ecosystem state.

Despite these difficulties, restoration projects must state goals in order to develop specific plans. The goal of "replacing what's been lost" thus gives way to more specific project objectives, which can be used to evaluate the success of restoration projects. In many cases, the process of restoration is mandated by a regulatory agency, who must determine whether or not a contract has been fulfilled. Several lessons follow from restoration planning in southern California wetlands.

Regional coordination, rather than piecemeal planning, is needed to maintain the salt marsh resources of the region. Within a region, the goal of salt marsh restoration should be to maintain populations of native species, by maintaining the natural variety of salt marsh habitats.[7] Specific objectives should focus on target habitat types and species. The solution to the problem is to define restoration goals within the region and then identify specific sites that have the greatest potential for achieving each objective. Regions should be broad enough to encompass the geographic distribution of several species.

The State Coastal Conservancy[8] adopted a two-county approach in assessing resources (eight remaining wetlands) and restoration goals near Los Angeles. The enhancement plan for Los Penasquitos Lagoon included a review of San Diego County resources for 14 remaining wetlands.[9] These local assessments help to insure that salt marsh populations are maintained where they originally occurred. For purposes of conserving the natural habitat diversity, however, assessments and plans should expand to include the southern California biogeographic region[10,11] from Point Conception south at least to the Mexico border. Because the wetland habitat remaining in the Los Angeles area is only 10% of historic acreages[8] and because local extinctions have occurred within individual wetlands, it will be necessary to draw upon a larger area to determine the likely character of natural habitats and to obtain species for reintroduction.

Off-site mitigation should be discouraged. Despite the fact that restoration planning should utilize broad geographic regions, mitigation for damages in one wetland should not take place at great distance. In planning for the Pacific-Texas oil pipeline, which would disturb large areas of fish habitats in Los Angeles Harbor, sites as far south as Tijuana Estuary (100 + miles away) were considered for the required 140-acre mitigation project. It is unlikely that regional resources can be maintained if large areas (such as the Los Angeles coastline) are developed even further without local mitigation. It is hard to imagine that damages to populations in the Los Angeles area could be compensated by modifying or creating wetland habitat in wetlands in San Diego County.

Species requirements must be well understood. Throughout the country, salt marsh vegetation types have been linked to specific intertidal elevations, soil salinities, and drainage conditions.[12-17] Yet the environmental conditions and species interactions that determine salt marsh community structure in southern California are only beginning to be discovered.[18,19] Algal mats beneath the salt marsh canopy have been investigated only in a few locations.[20-25] Likewise, the animal assemblages that are associated with the vascular plants are mostly unquantified. As Daiber[26] indicates, we know a lot about a few species (e.g., fiddler crabs and biting flies), but only a little about the vast array of insects, arachnids, and other arthropods, or even the more conspicuous mollusks and fishes that inhabit salt marshes. The biological interactions that structure salt marsh invertebrate communities have hardly been evaluated.[27] Unfortunately, we often learn about organism-habitat dependencies too late, when habitats decline to the point that dependent species become extinct.

Habitat for endangered species must be considered in developing regional restoration plans and site-specific objectives. Habitat destruction in southern California appears to be responsible for the endangerment of several wetland-dependent species. Those that are dependent on salt marsh habitats are the light-footed clapper rail (*Rallus longirostris levipes*), a large bird that is associated with the lower intertidal cordgrass (*Spartina foliosa*) community;[28] the Belding's Savannah sparrow (*Passerculus sandwichensis beldingi*), which depends on pickleweed (*Salicornia virginica*) for habitat;[29] and the salt marsh bird's beak (*Cordylanthus maritimus* spp. *maritimus*), an annual plant of the upper salt marsh. Restoration of habitat for endangered species is not only appropriate for the salt marsh of the region, it is mandated by U.S. Fish and Wildlife Service endangered species recovery plans.

Other species that are rare or uncommon should also be considered in setting restoration goals for southern California.[30,31] Two tiger beetles (*Cicindella trifasciata sigmoidea* and *C. gabbi*) occur in salt pannes of the upper marsh. The larvae of the wandering skipper (*Panoquina errans*) live only on saltgrass (*Distichlis spicata*), a plant of the salt marsh-sand dune transition. A plant at the northern limit of its geographic distribution (*Frankenia palmeri*) occurs only at one salt marsh in California. Because of tidal closure and coincident drought, two short-lived species were nearly eliminated at Tijuana Estuary.[19] The first, *Salicornia bigelovii*, occurs in only 7 of the 26 regional coastal wetlands; the second, *Suaeda esteroa*, occurs in only 14. Extinction at a single marsh would be a significant reduction in either species' distribution. All of the above-mentioned species should be considered as potential management targets in southern California.

Maintaining native plant communities that are uncommon within a region is also an appropriate restoration objective. For example, the transition from salt marsh to upland is nearly extinct in southern California because of urban and agricultural encroachment. The low marsh cordgrass community has disappeared from Los Penasquitos Lagoon due to altered tidal flushing.[32] Ditching to control mosquitoes has eliminated intertidal pools in many California salt marshes.[33-35] Brackish marshes adjacent salt marshes may have declined due to altered streamflow and lowered groundwater levels. Annual plants associated with alluvial fans are considered rare and in need of conservation.[36]

Reserves of rare plant species may need to be developed for artificial propagation and reintroduction to individual salt marshes following local extinction. Each of the species discussed above has unique requirements for establishment and persistence, and maintaining their populations within the region requires careful management. Several sites need to be identified for maintenance of each species, because none of the populations will be stable through time. Individual populations might face extinction if a natural or man-caused catastrophe alters its habitat. Gradually, the natural genetic diversity of the species will decline. The establishment of wetland gardens for rare marsh plants might prevent that trend.

Increasing "diversity" is not necessarily beneficial; maintenance of the "natural variety" of communities is a more appropriate regional goal. Maximizing or increasing diversity of species or habitat types is often cited as a site-specific goal of restoration. The implication is that "more is better." However, there are several reasons why this is inappropriate for salt marshes. First, many of the regional wetlands are dominated by pickleweed, which looks monotonous. From studies of the species, it is likely that monotypic stands of pickleweed are selected in lagoon marshes where environmental extremes are often prolonged. Long periods of hypersalinity and low soil moisture are tolerated by the perennial species of pickleweed, but not by many other salt marsh plants, e.g., cordgrass; annual pickleweed (*Salicornia bigelovii*); and sea blite (*Suaeda esteroa*).[19] Introduction of the latter into lagoon marshes is inadvisable.

Second, low diversity of vascular plants does not necessarily mean that the marsh as a whole lacks species. Associated with a monotypic pickleweed canopy can be a very rich

Table 1
SALINE MARSH SPECIES AND COMMUNITIES THAT
HAVE BEEN MENTIONED IN VARIOUS PROPOSALS AS
RESTORATION OBJECTIVES IN SOUTHERN CALIFORNIA

Target	Habitat
Light-footed clapper rail	Lower intertidal cordgrass
Belding's Savannah sparrow	Pickleweed marsh
Annual pickleweed	Middle intertidal marsh elevations
Sea-blite	Middle intertidal marsh elevations
Salt marsh bird's beak	Upper intertidal marsh
Mudflat tiger beetles	Salt pannes within the marsh
Wandering skipper	Salt grass, dune slack
Frankenis palmeri	Wetland-upland transition
Transitional community	Extreme high water and above
Intertidal pool fauna	Intertidal pools
Brackish marsh	Near streams or springs at the salt marsh edge

collection of epibenthic algae[25] and insects.[37] Too often, the suggestion that "we don't want just another pickleweed marsh" or "we want something more interesting" is based on a lack of understanding of the total ecosystem. In some cases, the diversity is there but not visible; in other cases, the diversity may be lacking because only a few species are suited to the site. Whenever alien species are encouraged to establish in favor of pickleweed, the potential value to Belding's Savannah sparrows may be jeopardized, in conflict with endangered species legislation. "More" is not always better.[7]

In summary, the regional goal for salt marsh restoration should be to maintain the *natural* diversity of species and community types. This can only be accomplished through coordinated planning for all wetlands in the region that are scheduled for restoration. A long list of specific objectives for individual sites (Table 1) should be considered, and the potential for each site to accommodate each objective should be evaluated. No single restoration site plan should necessarily include all potential objectives. Rather, there must be assurance at the regional scale that the natural biological resources will be maintained in perpetuity.

III. HYDROLOGICAL PLANNING

The most important factor that controls salt marsh composition is the hydrological regime. Levels of inundation, circulation, and salinity are all critical to plant and animal inhabitants. Equally important are the duration periods for each inundation, circulation, or salinity condition. Slight changes in the influence of either the tidal or riverine influxes can lead to major changes in the marsh community. Further restrictions on the hydrological situation are imposed by adjacent land uses. Returning areas to full tidal flushing will not be possible where roads or other structures have been placed below extreme high water. Thus, site plans for salt marsh restoration projects must be based on a careful review of existing and potential hydrological conditions. Several lessons follow from attempts to manipulate hydrology in southern California (Appendix 2).

Natural variations in hydrologic conditions must be maintained to manage for native salt marsh communities. The hydrology of the region is characterized by temporal variability, with seasonal, year-to-year, and occasional extremes of inundation, salinity, and circulation. There are basically two types of salt marshes in the region; those with pickleweed the overwhelming dominant, and those with a longer species list and abundant cordgrass in the lower marsh.[38] The former occur in lagoons, which are usually close to tidal flushing during the summer; the latter occur in estuaries with nearly continual access to tidal flushing. Both marsh types are subject to winter salinity dilution when streams flow, and to flooding when rainfall is heavy throughout the watershed. Catastrophic floods play a role that is disproportionate to their frequency.[32] Cordgrass expands its distribution substantially only during such unusual conditions.[39]

The duration of inundation is extremely important in controlling salt marsh composition.[19,38,40] Prolonged periods of high water are detrimental to many of the salt marsh vascular plants. Lower-marsh species, such as cordgrass, can tolerate prolonged inundation, but upper-marsh species, such as pickleweed, cannot.[41] Vegetation that is valued for its support of salt marsh-dependent endangered species may be eliminated if water is impounded for extended periods of time.

The winter period of rainfall and soil salinity reduction (the "low-salinity gap") must be brief to prevent a shift from saline to brackish marsh vegetation.[19] Prolonging the period of freshwater influx will shift the salt marsh composition toward brackish marsh vegetation. The brackish marsh species are establishment-limited; they are kept out of salt marshes by the persistence of saline-to-hypersaline soils. Brief periods of low-salinity in winter are insufficient to allow establishment of plants such as cattails (*Typha domingensis*), even though the mature plants are quite salt-tolerant.[42] Streamflow can be prolonged by reservoir discharge well after the winter flood season and by year-round release of treated wastewater; at such times, salt marsh soil salinities can be substantially diluted.[43-45] Reducing salinities for abnormally long periods allows germination and establishment of brackish marsh plants, causes major shifts in vegetation type, and leads to altered habitat values.

Prolonged periods of hypersaline drought will substantially alter salt marsh composition; thus, plans must provide for maintenance of tidal flushing during drought years. The 1984 growing season was entirely without rainfall throughout much of coastal southern California. At Tijuana Estuary, disturbances allowed closure of the estuary to tidal flushing. The barrier dune was trampled and destabilized. A major sea storm washed sand into channels and the tidal prism was reduced. Sand built up and eventually blocked the mouth. During 8 months of closure, a drought caused prolonged evaporation of channel and soil waters. Within a few months, salt marsh soils became three times as saline as seawater.[46] Heavy mortality of cordgrass occurred; two short-lived plant species neared extinction;[19] and marsh infauna were almost completely eliminated. The entire population of light-footed clapper rails (over 40 pairs) was lost that year, and recovery is not yet complete according to Paul Jorgensen, Estuarine Sanctuary Manager.

Maximizing tidal flushing is not an appropriate restoration goal, because extensive dredging has negative impacts. Because the maintenance of good tidal flushing has been widely accepted as a management goal for many of the regional lagoons and estuaries, several plans call for extensive dredging. While major dredging will reduce maintenance costs in systems subject to sedimentation, there can be significant negative impacts as well. In addition to excavating subtidal and intertidal ecosystems, deep dredging may eliminate wetland vegetation (as planned for Batiquitos Lagoon) and accelerate slumping and erosion of salt marsh areas upstream (as appears to have occurred at Elkhorn Slough). Areas that are poorly flushed may have inherent values that will be altered by increased circulation. For example, where cordgrass and pickleweed compete for space and for soil nitrogen supplies,[18] the vegetation would probably shift toward pickleweed dominance under conditions of better drainage and toward cordgrass when water tends to impound. This is because pickleweed is at a disadvantage where drainage is sluggish, while cordgrass is at an advantage. If the area is being managed for light-footed clapper rails, maximum tidal flushing may be almost as detrimental as minimal tidal flushing.

In summary, there are no simple hydrological goals, such as "increase tidal flushing" that can be promoted. Careful study of existing hydrology and constraints on changing fresh/seawater influence are needed before the range of potential restoration projects can be determined. Then, the regimes (timing and duration) must be specified for inundation, circulation, and salinity required to support the desired salt marsh community.

IV. EXPERIMENTATION

Salt marsh restoration is still a trial-and-error process, although it is rarely viewed as such. Instead of recognizing that we are still plotting the unknown, proposers of restoration projects and especially proponents of restoration as mitigation for development make staggering promises. For example, Kel-Cal[47] claimed that their project in Agua Hedionda Lagoon would have "major environmental benefits," that it "increases quality and diversity of salt marsh food chain through enhanced tidal flushing, nutrient exchange and maintenance of water salinity levels" and "enhances functional capacity of overall lagoon ecosystem" and "increases the quantity, quality and diversity of wetlands habitat and related fish, wildlife and aesthetic values." Every project that has been implemented in California appears to have had some type of problem (cf. review of Race and Christie[48] for San Francisco Bay wetlands). There are no guarantees that projects will proceed as planned or that plans are adequate to begin with (Appendix 1).

The responsibility of federal agencies is to insure "no net loss of in-kind habitat" (U.S. FWS policy) in mitigation planning. Yet the state-of-the-art is not sufficient to insure that 100 or even 200 acres of restored habitat will mitigate losses caused by the destruction of 100 acres of functional salt marsh somewhere else. Mitigation funds are becoming available at a rapid pace in southern California, where projects such as the Pacific-Texas oil pipeline could make possible the restoration of >100 acres of wetland at a time.

It is necessary to move salt marsh restoration away from its present trial-and-error approach and into the realm of ecotechnology — the careful manipulation of ecosystems to achieve desired management goals. To do this requires a scientific approach, specifically, the use of controlled, replicated, field experiments to determine the best procedures for controlling species composition and ecosystem functioning.[49] Then, careful evaluation of experimental results will make it possible to record not only what worked but reasonable explanations of *why* it worked. It will also be possible to predict whether those restoration measures are likely to be successful at other sites. Likewise, knowledge of failures and the causes of failure will make it possible to avoid their repetition.

Every opportunity to add an experimental element to restoration plans needs to be followed. Many restoration projects encompass several acres of habitat, and at least a small portion of such sites could be designated for experimental work. Any objectives that are not certain to be implementable at the site or for which there are alternative methods for reaching the objectives can be set up as treatments and appropriately evaluated. For example, the dredge spoil island at the Port of San Diego was to be planted with cordgrass, with no previous experience to guide planting requirements. Research on the smooth cordgrass (*Spartina alterniflora*) of eastern salt marshes identified elevation and salinity as important controlling factors,[50,51] and local research suggested that Pacific cordgrass had limited tolerance to high intertidal conditions and to extreme hypersalinity.[7] Questions that remained were (1) would cordgrass grow in the extremely saline sediments (60 ppt and above), (2) what maximum and minimum elevations would support cordgrass, (3) was soil augmentation (adding sand) required to improve drainage, and (4) how close should stock be planted?

Prior to planting the entire 40-acre site, an experiment was conducted, varying soil salinity (by locating 125 plantings along 10 transects throughout the site), elevation (plantings at 4.5, 5.0, and 5.5 ft MLLW [mean lower low water]), and soil augmentation (rows with and without sand added to holes dug for the plants). Success was determined through monthly monitoring of survival and expansion from the April 1984 planting date to December 5, 1984.[52] The experiment documented that the site with lowest salinity produced the most vigorous plants and had 98.5% survival. The most robust plants were at 4 to 5 ft above MLLW; substantial vegetative expansion occurred; soil augmentation did not alter survival or vigor. This simple experiment demonstrated that the site was suitable for cordgrass and it led to a simpler planting scheme, by showing that 12-ft planting intervals were adequate and that soil augmentation was not required. Furthermore, it established a vigorous nursery of plants for future transplanting. Unfortunately, monitoring of survival alone could not tell us why the vegetation was able to grow so well.

Both large and small restoration projects can be used to experimental advantage. Projects that encompass hundreds of acres often have the potential to develop complicated engineering designs. This makes it possible to test several ideas about the influence of hydrology on coastal ecosystems, e.g., the degree and timing of tidal flushing needed for different salt marsh communities. Even small projects, such as the proposal to create 8500 ft² of intertidal habitat in Tijuana Estuary in exchange for habitat losses in San Diego Bay, can include experiments to assess the degree of tidal flushing necessary to maintain pickleweed or cordgrass vegetation.

Funding should be available from federal agencies to support experiments that will improve restoration projects. The above recommendations may appear idealistic; however, there are opportunities to implement them. In its newly formulated Wetland Research Plan,[53] the Environmental Protection Agency (EPA) proposed that major wetland restoration projects be considered for experimentation that would lead to improved mitigation planning. The EPA involvement in mitigation projects comes through their responsibilities under Section 404 of the Clean Water Act. As mitigation proposals are evaluated, the EPA has the opportunity to require modifications that include the engineering designs for appropriate experiments. This can be done through consultation with the scientific community, at which time, scientists would become aware that sites are available for field studies. The EPA would then be a suitable agency to fund the research portion of projects including experimental work. Results from the work would then be funneled directly to agencies responsible for mitigation planning.

In summary, the ability to create, enhance, or restore specific salt marsh characteristics requires a science of ecosystem management, or ecotechnology, based on hypothesis testing through field experimentation.

V. ASSESSMENT OF RESTORATION SUCCESS OR FAILURE

There are at least two major reasons why careful assessment of restoration projects is necessary. First, resource agencies need to keep track of how much of the regional wetland communities are being restored. Second, in the case of mitigation trade offs, regulatory agencies need to determine if project proponents have fulfilled their requirements. The lessons that emerge from California concern not only the criteria used to judge success, but also the procedures used to obtain useful data and to interpret those results.

Two criteria must be considered in judging project success: (1) comparison of what actually developed with what was promised and (2) comparison of what developed with what was and is still needed in the region. In order to judge whether a mitigation project has been completed, the attributes of the restored salt marsh must be compared to its specific stated objectives. Most salt marsh restoration projects fail by this measure,[48] even if the project has produced a vegetation type that is appropriate for the region. For instance, Muzzi Marsh in the San Francisco Bay area was planned for cordgrass; the plantings failed but a robust stand of pickleweed developed in its place.[54] The project was a failure but a salt marsh typical of the region did develop. The alternative criterion is how well the project fulfilled region-wide restoration goals. Some have argued that pickleweed marshes are more appropriate for the Bay area; thus, the project was successful, and what failed was the planning process.

If site-specific objectives are set on the basis of regional needs, individual projects must be expected to fulfill their role, or some regional resources will be lost. Not all projects that produce pickleweed-dominated vegetation instead of an alternative community type can be accepted and still conserve the diversity of salt marsh habitats that characterize the region. However, careful thought must also be given to the values of unplanned events that may follow project implementation. If a project is planned to provide marsh and fishery habitat but first develops a nesting colony of least terns (as at the dredge spoil island at the Port of San Diego), it may contribute unexpectedly to regional restoration goals. Such "happy accidents" must be acknowledged and initial requirements reevaluated. Plans for other restoration projects might need to be reevaluated as well. Both criteria should be used in an overall evaluation of individual projects. Thus, a broadly based, permanently staffed data center is needed to record restoration accomplishments.

Assessment of the marsh that develops on a restoration site is a long-term process. While pickleweed can usually be expected to invade and dominate a site within a year, establishment of cordgrass and expansion to required densities may require several years.[50] Establishment of breeding populations of target endangered species, e.g., Belding's Savannah sparrows and light-footed clapper rails, may take decades. However, most project evaluations are needed within about a year of implementation. The establishment of cordgrass at the dredge spoil island of the Port of San Diego will be evaluated after only 2 weeks, at which time 90% of the transplants must still be alive. Chances of "success" are high, as it is hard to tell if wilted plants are actually dead within that time frame. Such short time intervals for assessment of project success are inappropriate.

Assessing ecological communities requires detailed and frequent sampling. Moss[55] acknowledges that his understanding of how disturbances have altered marshes of the Broads, Norfolk, England, is based on a 10-year research program.

Data sets and their interpretation must be able to stand the test of peer review. A major controversy developed over the ecological relationship of habitats on Gunpowder Point (a 40-acre island of upland habitat surrounded by wetlands along the edge of San Diego Bay) and adjacent wetlands when the Point was proposed for development (a 7-story 700-room hotel). Proponents said that there were no significant links between the two habitats. Opponents expressed concern that hotel construction would have negative impacts on endangered salt marsh birds (Belding's Savannah sparrow and light-footed clapper rails). There was particular concern that the island uplands provided the only refuge for marsh birds during periods of highest water. On the basis of this and other biological issues, the project was denied by the California Coastal Commission.[56,57]

Funds were then provided to the project proponents for a study to resolve the biological issues and a consultant was selected to evaluate the ecological linkages between the upland and the marsh. The consultants[58] based that evaluation on (1) hypothetical food webs for the upland and the marsh and (2) sparrow foraging observations for short periods during May, July, October, and December 1982 (a total of 82 person-hours). The consultants concluded that interactions between the upland and wetland were minimal. At the public hearings for the revised development project, the report failed to satisfy peer reviewers; yet it was the only available data set. The Commission approved the plans to construct an 8- to 12-story hotel and visitor center on Gunpowder Point. Restoration measures were required to mitigate the impacts of development, along with monitoring of their success. At present, the proponents are designing a monitoring program.

At issue here is the process of ecological assessment and monitoring. The proponents of a project are usually the ones required to assess the impacts of their project or to monitor the changes following their restoration practices. They have a vested interest in showing no impact of development or successful mitigation. Consultants to developers have a strong interest in pleasing their clients. Opponents of development projects do not receive funding to refute claims or demonstrate inadequacy of data sets. Objective, peer-reviewed science usually plays a minor role in the evaluation process, unless a previously published paper bears on the topic. Ecological sampling is rarely carried out with a regional or long-term perspective on how wetlands function. Data gathered at one project site will not necessarily be comparable with those elsewhere, because sampling methods may differ. Data from a short-term monitoring program, even with a comprehensive sampling program, will probably be inadequate to interpret cause-effect relationships, because results must be interpreted in light of long-term variability in environmental conditions. The result is very slow progress in understanding how individual projects affect wetlands, whether restoration projects meet their objectives, or why restoration succeeds or fails. Again, a regional data center is needed to oversee projects from a broad perspective.

Monitoring programs should be open to scientific review and responsive to criticism. There are a number of potential problems with ecological studies done under contract to developers. First, the decision to select monitors for the project may be based on cost or number of stations to be monitored, rather than the ability to evaluate results within the context of regional wetland dynamics. Second, the sampling locations or frequencies may not be adequate to determine cause-effect relationships. Given the year-to-year dynamics of Pacific Coast marshes, sampling programs have to be flexible to capture the system at critical times. Third, results must be evaluated in relation to temporal patterns of years past. A cumulative data set for several wetlands must be developed, with each site evaluated in relation to the regional data base. Patterns that cannot be interpreted on the basis of the data for 1 year or one sampling site may well be understandable when viewed with a regional perspective. Fourth, results and their interpretation must be scrutinized by scientific peers in order for the nonscientific audience to have confidence in their efficacy. Rarely are project reports circulated for peer review; even after being filed with the regulatory agency, such reports are difficult to obtain. There needs to be a mechanism for evaluating monitoring programs and for making their faults or fine points known.

In summary, piecemeal assessment will be inadequate and very likely misleading. What is needed is a regional scientific assessment that is long-term in nature, open to peer review, and subject to continued improvement. It is a function that is appropriate for university or government research laboratories, but rarely compatible with the constraints placed on individual contracts. The requirements are stability of resources to support assessment (such as long-term storage of large data sets), the ability to alter plans and respond to rare events that occur at a site (i.e., to increase sampling even if an individual budget does not allow it), and the commitment to understand regional patterns (not just sample a site). None of these requirements can be fulfilled by individual, short-term assessments. The collective experience of a research program is required, along with the subsidies provided by university or government agencies in their fostering of research and publication. The assessment of regional restoration efforts must be a scientific function if salt marsh restoration is to move into the realm of science.

Appendix 1

SAMPLE OF SALT MARSH RESTORATION PROJECTS IN SOUTHERN CALIFORNIA

Each of the following examples is a greatly shortened description of a large project. Some components of each may have been properly planned; however, the problems are emphasized to provide lessons for improvement.

Kel-Cal[47] proposed to mitigate roadfill on 13.6 acres of wetland at Agua Hedionda Lagoon by modifying existing habitat in two ways. First, deposition of dredge spoil on top of salt marsh habitat (including pickleweed) was proposed to create a 2-acre "bird-nesting island". The area was judged too small for use by the endangered California least tern; it would function as a dredge disposal site more than as mitigation for road fill. Second, "brackish ponds" were planned to be dug in areas of disturbed transitional habitats, where marsh graded into upland. As a rare habitat within the region, the transition habitat was appropriate for restoration by removal of exotic species and reestablishment of natives. Instead, however, the project sought to excavate depressions that would collect brackish water. In implementing the project, pits were dug to 2-m depth, and still did not encounter groundwater. A potentially useful restoration site was substantially modified and reduced in habitat quality; yet the alterations were considered mitigation for roadfill.

The San Diego Unified Port District[59] was allowed to discharge dredge spoils from the J-Street Boat Basin in Chula Vista into south San Diego Bay on the condition that they develop the dredge spoil island into a wildlife reserve. However, the spoils failed to dewater on schedule, and planting of marsh vegetation has been delayed years beyond the planned completion date. Each year since 1984, access to the site has been restricted due to the California least tern, an endangered species that is protected from disturbance during its nesting season, from arrival in March or April through departure in August. A pilot planting of cordgrass was accomplished in 1984, with full-scale planting planned for 1985. However, there were contractual difficulties in 1985 that were not resolved prior to arrival of least terns. Implementation finally occurred in spring 1986. Further complications of the project included the failure to provide the fisheries habitat in the original site design, and channels must still be created to satisfy mitigation requirements. Cordgrass transplantation at this site has been highly successful, and utilization of higher elevations by least terns has been an added bonus for the site.

Like most restoration projects in the region, this one represents a change in type, rather than a net gain, of wetland habitat, because the island replaced a large subtidal area in south San Diego Bay. In addition, there was a substantial period when no habitat was present — during construction and dewatering of the island. The extent to which damages caused by dredging at J-Street Boat Basin were mitigated is thus very difficult to quantify. At least until the entire dredge spoil island is suitable for native communities, the project has reduced the area available for maintenance of natural resources.

The California Department of Transportation (Caltrans) proposed to widen Freeway 5 and to construct a freeway interchange with Highway 54 in an area that would partially cover the Paradise Creek salt marsh in San Diego Bay. Construction of a flood control channel on the Sweetwater River was included in the plans, with the channel to be built between the east- and west-bound portions of Highway 54. Among the mitigation measures proposed was conversion of approximately 25 acres of disturbed upper marsh into lower marsh habitat for the light-footed clapper rail and least tern. As in the Agua Hedionda Lagoon project, areas of upper marsh and transition to upland were to be traded for lower marsh and channels. The creation of habitat for endangered birds was credited as mitigation, at the expense of another diminishing habitat type which, in itself, was worthy of restoration.

At least one new problem occurred during project implementation. Sediments that were high in lead content were encountered during the removal of soils from the disturbed upper marsh. The lead was thought to have originated from paints that were discarded at a time when wetlands were used as urban dumps. Soil from the contaminated area had to be trucked to a toxic waste dump before the required low-marsh habitat could be created, according to J. Gidley of Caltrans. The potential for mobilization of toxic materials must be considered in restoration projects. The U.S. Army Corps of Engineers recognized this potential in building wetlands from dredge spoils; the Corps included study of heavy metals in its Dredged Material Research Program.[60-62]

Several projects have been proposed to remove sediments that accumulate in the Southern California coastal lagoons, as highlighted in the *Los Angeles Times,* March 31, 1985.

The California Fish and Game Department had the eastern portion of Buena Vista Lagoon dredged to provide deeper aquatic habitat. Spoils were deposited alongside the dredged area within the lagoon. It was expected that marsh vegetation would develop on some of the spoils. It has not, and there was little reason to expect rapid invasion by salt marsh plants. High evaporation rates cause hypersalinity in exposed dredge spoils, and soil structural changes result in brick-like clods with drying. Samples of the exposed sediments were analyzed in 1985, and soil paste conductivities indicated that salinities were several times that of seawater.

The California Department of Fish and Game also proposed the creation of several ponds from existing marsh habitat at Pismo Lake[63] where the one-time shallow lake had gradually filled in and developed into a marsh. Again, the spoils were to be dumped within the wetland, with the expectation that vegetation would invade. The planned openwater habitats would encompass 8.4 acres; the spoil deposits would create 2.9 acres of islands. If the upland islands would not become vegetated, there would be a net loss of habitat and certainly a net loss of wetland habitat, as existing areas of saline-brackish marsh were covered with spoil. The Pismo Lake project was termed wetland restoration because gradually accumulated sediments were to be removed; however, the net effect was to change one type of wetland (brackish and saline marsh) into another (ponds). Whether or not this represents an increase or decrease in habitat quality depends on the value system.

Natural sedimentation processes are clearly accelerated by disturbance and erosion within coastal watersheds; however, arguments for reversing the process must be weighed against arguments for protecting the habitat values of accreting marshes. Where spoils cannot be disposed of off-site, the effect of dredging is detrimental because of the net loss in wetland habitat.

The State Coastal Conservancy sponsored a transition-zone restoration project at Tijuana Estuary that sought to (a) eliminate exotic weeds (especially *Chrysanthemum*) on an upland transition habitat and (b) replace the disturbed area with native species. Despite recommendations not to disturb the soil, the site was disked and sown with native species. *Chrysanthemum* quickly reestablished, as expected of weedy species. Native seedlings did not thrive. The project failed to achieve either objective.

Appendix 2

PROBLEMS WITH HYDROLOGICAL PLANS FOR CALIFORNIA WETLAND RESTORATION PROJECTS

Early plans for the restoration of Batiquitos Lagoon called for the stabilization of water levels with a wier. Biological consultants claimed that salt marsh vegetation would be unaffected by the shift from a tidal to a nontidal hydrological regime. These claims were contrary to the ecological understanding of controls on species composition.[19,32,38] Limited sampling for depth to groundwater was inadequate to determine the role of water-level oscillations in controlling vegetation structure.

A reanalysis of the restoration objectives led to plans for substantial dredging to achieve full tidal flushing. However, a controversy developed over the volume of dredging required to maintain an open mouth to the lagoon, and an even more ambitious dredging plan was adopted to reduce dredging maintenance costs. As a result, areas that are now occupied by salt marsh vegetation will be destroyed. The promise of reduced maintenance costs took precedence over the regional restoration goal of maintaining salt marsh habitat, according to P. Williams, Hydrologist, and L. Marcus, State Coastal Conservancy.

Previously diked areas of Elkhorn Slough in central California were known to have subsided following the exclusion of tides. When funds became available to restore a pasture to intertidal salt marsh, it was clear that the topography would have to be recontoured. Thus, plans were developed to cut channels and use the spoils to build islands of marsh habitat. However, after a long series of problems in implementing the project, bids were solicited, and one contractor promised to move a substantially larger volume of spoil than required. The California Department of Fish and Game accepted that bid, although it did not conform with project plans. Channels were dug in a herringbone pattern with spoils heaped alongside. The marina-like configuration was not appropriate to the goal of habitat restoration of the National Estuarine Sanctuary.

Preliminary plans for restoration of Ballona Wetland called for augmenting tidal flow into an area dominated by pickleweed marsh. At present, culverts restrict tidal flows into the salt marsh. The reason for increasing flows was not to alter the pickleweed marsh but to extend tidal influence to disturbed marsh habitats further upstream. Technical advisors to the planning group studied the elevations and predicted that increased tidal flow would eliminate pickleweed by increasing inundation periods beyond its tolerance.

An experiment was performed to test the model of tidal flow. Tide gates were opened, on-site observations of water levels were made, and the area was photographed by plane to assess hydrological conditions. The resulting information substantiated concerns that increased flushing would damage pickleweed. An alternative plan was developed with a second access to tidal water to be created for the inner marsh area. Thus, new marsh habitats could be created and the existing pickleweed marsh could be preserved.[64]

Prolonged discharge of water from reservoirs is usually done without regard for impacts on downstream wetlands. In the San Diego River, reservoir discharge continued for months beyond the 1980 flood period. Excessive freshwater flooding of the intertidal salt marsh killed pickleweed, reduced soil salinity for several months, and allowed cattails (*Typha domingensis*) to invade and dominate the marsh.[19,40,42]

REFERENCES

1. **Bradshaw, A. D. and Chadwick, M. J.,** *The Restoration of Land,* Blackwell Scientific, Boston, 1980.
2. **Harvey, H. T., Williams, P., Haltiner, J.,** Philip Williams and Associates, Madrone Associates, and Staff of San Francisco Bay Conservation and Development Commission, Guidelines for enhancement and restoration of diked historic baylands, Report prepared for San Francisco Bay Conservation and Development Commission, 1983.
3. **Josselyn, M., Ed.,** Wetland restoration and enhancement in California, A California Sea Grant Program Publication, Report No. T-CSGCP-007, La Jolla, Calif., 1982.
4. **Josselyn, M. and Buchholz, J.,** Marsh restoration in San Francisco Bay: a guide to design and planning, Paul F. Romberg Tiburon Center for Environmental Studies, Tiburon, Calif., 1984.
5. **Lewis, R. R., III, Ed.,** *Creation and Restoration of Coastal Plant Communities,* CRC Press, Boca Raton, Fla., 1982.
6. **Pullen, E. J., Yancey, R. M., Knutson, P. L., and Hurme, A. K.,** An annotated bibliography of CERC Coastal Ecology Research, U.S. Army Corps of Engineers, Fort Belvoir, Va., 1980.
7. **Zedler, J. B.,** Salt marsh restoration: a guidebook for southern California, California Sea Grant Report No. T-CSGCP-009, 1984.
8. State Coastal Conservancy, Regional wetland restoration study: Los Angeles and Orange Counties final draft report, 1982.
9. Los Penasquitos Lagoon Foundation and the State Coastal Conservancy, Los Penasquitos Lagoon enhancement plan and program, State Coastal Conservancy, 1985.
10. **Littler, M. M.,** Overview of the rocky intertidal systems of southern California, in *The California Islands: Proceedings of a Multidisciplinary Symposium,* Power, D. M., Ed., Santa Barbara Museum of Natural History, Santa Barbara, Calif., 1980, 265.
11. **Seapy, R. R. and Littler, M. M.,** Biogeography of rocky intertidal macroinvertebrates of the southern California Islands, in *The California Islands: Proceedings of a Multidisciplinary Symposium,* Power, D. M., Ed., Santa Barbara Museum of Natural History, Santa Barbara, Calif., 1980, 307.
12. **Cowardin, L. M., Carter, V., Golet, F. C., and LaRoe, E. T.,** Classification of wetlands and deepwater habitats of the United States, Office of Biological Services, Fish and Wildlife Service, U.S. Department of the Interior, FWS/OBS-79/31, Washington, D.C., 1979, 103.
13. **Josselyn, M.,** The ecology of San Francisco Bay tidal marshes: a community profile, U.S. Fish and Wildlife Service, Division of Biological Services, FWS/OBS-83/23, Washington, D.C., 1983, 102.
14. **Nixon, S. W.,** The ecology of New England high salt marshes: a community profile, U.S. Fish and Wildlife Service, Office of Biological Services, FWS/OBS-81/55, Washington, D.C., 1982, 70.
15. **Onuf, C. P.,** The ecology of Mugu Lagoon: an estuarine profile, U.S. Fish and Wildlife Service, Office of Biological Services, Biol. Rep. 85(7.15), Washington, D.C., 1987, 122.
16. **Pomeroy, L. R. and Weigert, R. G., Eds.,** *The Ecology of a Salt Marsh,* Springer-Verlag, New York, 1981.
17. **Seliskar, D. M. and Gallagher, J. L.,** The ecology of tidal marshes of the Pacific Northwest Coast: a community profile, U.S. Fish and Wildlife Service, Division of Biological Services, FWS/OBS-82/32, Washington, D.C., 1983, 65.
18. **Covin, J.,** The Role of Inorganic Nitrogen in the Growth and Distribution of *Spartina foliosa* at Tijuana Estuary, California, M.S. thesis, San Diego State University, Calif., 1984, 60.
19. **Zedler, J. B. and Beare, P. A.,** in *Estuarine Variability,* Wolfe, D., Ed., Academic Press, San Diego, 1986, 295.
20. **Sage, W. W. and Sullivan, M. J.,** Distribution of bluegreen algae in a Mississippi Gulf Coast salt marsh, *J. Phycol.,* 14, 333, 1978.
21. **Sullivan, M. J.,** Edaphic diatom communities associated with *Spartina alterniflora* and *S. patens* in New Jersey, *Hydrobiologia,* 52, 207, 1977.
22. **Sullivan, M. J.,** Diatom community structure: taxonomic and statistical analyses of a Mississippi salt marsh, *J. Phycol.,* 14, 468, 1978.
23. **Sullivan, M. J.,** Distribution of edaphic diatoms in a Mississippi salt marsh: a canonical correlation analysis, *J. Phycol.,* 18, 130, 1982.
24. **Williams, R. B.,** The Ecology of Diatom Populations in a Georgia Salt Marsh, Ph.D. thesis, Harvard University, Cambridge, Mass., 1962.
25. **Zedler, J. B.,** Salt marsh algal mat composition: spatial and temporal comparisons, *Bull. South. Calif. Acad. Sci.,* 81, 41, 1982.
26. **Daiber, F. C.,** *Animals of the Tidal Marsh,* Van Nostrand Reinhold, New York, 1982.
27. **Kneib, R. T.,** Patterns of invertebrate distribution and abundance in the intertidal salt marsh: causes and questions, *Estuaries,* 7, 392, 1984.
28. **Jorgensen, P. D.,** Habitat Preference of the Light-Footed Clapper Rail in Tijuana Marsh, California, M.S. thesis, San Diego State University, Calif., 1975.

29. **Massey, B. W.,** Belding's Savannah sparrow, Southern California Ocean Studies Consortium, California State University Contract No. DACW09-78-C-0008, U.S. Army Corps of Engineers, Los Angeles District.
30. **Nagano, C. D.,** California coastal insects: another vanishing community, *Terra,* 19, 27, 1981.
31. **Powell, J. A.,** Endangered habitats for insects: California coastal sand dunes, *Atala,* 6, 41, 1978.
32. **Zedler, J. B.,** Catastrophic flooding and distributional patterns of Pacific cordgrass (*Spartina foliosa* Trin.), *Bull. South. Calif. Acad. Sci.,* 74, 1986.
33. **Balling, S. S. and Resh, V. H.,** Arthropod community response to mosquito control recirculation ditches in San Francisco Bay salt marshes, *Environ. Entomol.,* 11, 801, 1982.
34. **Barnby, M. A., Collins, J. N., and Resh, V. H.,** Aquatic macroinvertebrate communities of natural and ditched potholes in a San Francisco Bay salt marsh, *Estuarine, Coastal Shelf Sci.,* 20, 331, 1985.
35. **Resh, V. H. and Balling, S. S.,** Ecological impact of mosquito control recirculation ditches on San Francisco Bay marshlands: study conclusions and management recommendations. in Proc. and Papers 51st Annu. Conf. California Mosquito and Vector Control Assoc., Inc., 1983, 49.
36. **Ferren, W. R., Jr.,** Carpinteria salt marsh. Environment, history, and botanical resources of a southern California estuary, Department of Biological Sciences, University of California, Santa Barabara, 1985.
37. **Nagano, C. D., Hogue, C. L., Snelling, R. R., and Donahue, J. P.,** The insects and related terrestrial arthropods of Ballona, in *The Biota of the Ballona Region, Los Angeles County,* Schreiber, R. W., Ed., 1981.
38. **Zedler, J. B.,** The ecology of southern California coastal salt marshes: a community profile, U.S. Fish and Wildlife Service, FWS/OBS-81/54, Washington, D.C., 1982, 110.
39. **Zedler, J. B.,** Freshwater impacts in normally hypersaline marshes, *Estuaries,* 6, 346, 1983.
40. **Zedler, J. B.,** The San Diego River marsh before and after the 1980 flood, *Environment Southwest,* 495, 20, 1981.
41. **Mahall, B. E. and Park, R. B.,** The ecotone between *Spartina foliosa* Trin. and *Salicornia virginica* L. in salt marshes of northern San Francisco Bay. III. Soil aeration and tidal immersion, *J. Ecol.,* 64, 811, 1976.
42. **Beare, P. A.,** Salinity Tolerance in Cattails (*Typha domingensis* pers.): Explanations for Invasion and Persistence in a Coastal Marsh, M.S. thesis, San Diego State University, Calif., 1984, 57.
43. **Zedler, J. B., Koenigs, R., and Magdych, W. P.,** Streamflow for the San Diego and Tijuana Rivers, San Diego Association of Governments, San Diego, 1984.
44. **Zedler, J. B., Koenigs, R., and Magdych, W. P.,** Review of salinity and predictions of estuarine responses to lowered salinity, State of California Water Resources Control Board, San Diego Association of Governments, San Diego, 1984.
45. **Zedler, J. B., Magdych, W. P., and San Diego Association of Governments,** Freshwater release and southern California coastal wetlands: management plan for the beneficial use of treated wastewater in the Tijuana River and San Diego River estuaries, San Diego Association of Governments, San Diego, 1984.
46. **Zedler, J. B. and Nordby, C. S.,** The ecology of Tijuana Estuary: an estuarine profile, U.S. Fish and Wildlife Service, Washington, D.C., in press.
47. **Kel-Cal,** Public Notice/Application No. 85-137-AA, U.S. Army Corps of Engineers, Ft. Belvoir, Va., 1985.
48. **Race, M. and Christie, D.,** Mitigation, marsh creation and decision-making in the coastal zone, *Environ. Manage.,* 6, 317, 1982.
49. **Zedler, J. B.,** Salt marsh restoration: the experimental approach, in *Coastal Zone '83,* Magoon, O. T., Ed., American Society of Civil Engineers, New York, 1983, 2578.
50. **Broome, S. W., Seneca, E. D., and Woodhouse, W. W., Jr.,** Long-term growth and development of transplants of the salt marsh grass *Spartina alterniflora, Estuaries,* in press.
51. **Woodhouse, W. W., Jr., and Knutson, P. L.,** Atlantic coastal marshes, in *Creation and Restoration of Coastal Plant Communities,* Lewis, R. R., III, Ed., CRC Press, Boca Raton, Fla., 1982, 45.
52. **Winfield, T. P.,** Chula Vista wildlife reserve transplant program, Woodward-Clyde Consultants, Walnut Creek, Calif., 1985.
53. **Zedler, J. B. and Kentula, M. E.,** Wetlands research plan, November, 1985. Environmental Research Laboratory, U.S. Environmental Protection Agency, Corvallis, Ore., 1985.
54. **Faber, P.,** Marsh restoration with natural revegetation: a case study in San Francisco Bay, in *Coastal Zone '83,* Magoon, O. T., Ed., American Society of Civil Engineers, New York, 1983, 729.
55. **Moss, B.,** The Norfolk Broadland: experiments in the restoration of a complex wetland, *Biol. Rev.,* 58, 521, 1983.
56. **Metz, E. D. and Zedler, J. B.,** Who says science can't influence decision making?, in *Coastal Zone '83,* Magoon, O. T., Ed., American Society of Civil Engineers, New York, 1983, 584.
57. **Metz, E. D. and Zedler, J. B.,** Using science for decision making: the Chula Vista bayfront local coastal program, *Environ. Impact Assess. Rev.,* 4, 250, 1983.

58. **Jones and Stokes Associates, Inc.,** Analysis of select biological issues relating to the Chula Vista Bayfront plan, Prepared for the City of Chula Vista, Calif., 1983.

59. San Diego Unified Port District, Chula Vista boat/basin wildlife reserve, draft EIR, 1975.

60. Center for Wetland Resources, Trace and toxic metal uptake by marsh plants as affected by Eh, pH, and salinity, Dredged material research program, U.S. Army Corps of Engineers, Waterways Experiment Station, Vicksburg, Miss., 1977.

61. **Gunnison, D.,** Mineral cycling in salt marsh-estuarine ecosystems, Dredged material research program, U.S. Army Corps of Engineers, Waterways Experiment Station, Vicksburg, Miss., 1978.

62. **Lee, C. R., Smart, R. M., Sturgis, T. C., Gordon, R. N., Sr., and Landin, M. C.,** Prediction of heavy metal uptake by marsh plants based on chemical extraction of heavy metals from dredged material, Dredged material research program, U.S. Army Corps of Engineers, Waterways Experiment Station, Vicksburg, Miss., 1978.

63. California Coastal Commission Permit Request #4-85-221, Central Coast District, Santa Cruz, Calif., 1985.

64. **Shapiro and Associates,** Enhancement, restoration, and management plan for the Ballona Wetland, The National Audubon Society, Sacramento, Calif., 1985.

Chapter 9

IMPROVING COAL SURFACE MINE RECLAMATION IN THE CENTRAL APPALACHIAN REGION

W. Lee Daniels and Carl E. Zipper

TABLE OF CONTENTS

I.   Introduction 140
        A.   The Powell River Project 140
        B.   Problems in Rehabilitating Surface Mines in the Central
                Appalachians 143

II.   Controlled Overburden Placement 143
        A.   Use of Natural Topsoils vs. Topsoil Substitutes 143
        B.   Selecting Geologic Materials for Use as Topsoil Substitutes 144
        C.   Mine Soil Construction to Insure Reclamation Success 145

III.   Establishing and Managing Vegetation 147
        A.   Revegetation Strategies and SMCRA 147
        B.   Initial Establishment of Vegetation 147
        C.   Factors Critical to Long-Term Revegetation Success 148

IV.   Using Mining Spoil to Construct Postmining Landforms 150
        A.   Application of Controlled Overburden Placement Procedures on Lands
                Returned to AOC 151
        B.   AOC Backfill Stability 151
        C.   An Alternative to AOC 152
        D.   Comparative Economics of Landform Alternatives 155

V.   Institutional Constraints to Improved Reclamation 158

VI.   Conclusions 160

References 161

I. INTRODUCTION

Large areas of steeply sloping land in the Appalachian Plateau region of the U.S. are underlain by high grade coal deposits. Many of these coal seams are shallow enough to be mined by surface techniques, and since the turn of the century, hundreds of thousands of hectares have been disturbed by surface mining activities. Early mining, particularly that performed before the 1970s, often resulted in severe environmental degradation. Acid mine drainage, barren unvegetated mined areas, and steep unstable slopes of mining spoil were frequently left behind after mining. Lax (or nonexistent) state surface mining regulations gave no legal or economic incentives for the mining industry to adequately reclaim these lands. While many of these areas were successfully revegetated, the postmining land use potential of the resultant highwall-bench-outslope topography (Figure 1) was often limited by spoil compaction and instability, narrow benches, and poor access. Many reclamation activities were simply not cost-effective to the operator who was driven to produce coal in the most efficient possible manner.

Many states began to enforce more stringent environmental regulations in the 1960s and 1970s, and the quality of the postmining environment began to improve. In 1977, Congress passed the Surface Mining Control and Reclamation Act (SMCRA) which required each state to develop and enforce its own Permanent Regulatory Program for surface mining. Strict performance standards were set for all aspects of exploration, permitting, mining, and reclamation. The major goals of the environmental protection standards were long-term revegetation and erosion control, and protection of surface water quality. To meet these goals, miners are required to post a large bond for each mined area and to meet performance standards for a minimum period of 5 years after mining.

Unfortunately, many of the mining and reclamation techniques required by SMCRA were essentially unknown to large segments of the mining industry. The entire approach to surface mining had to change to accommodate environmental concerns, and many aspects of the law have been bitterly opposed. In the Appalachian region for example, the provision requiring the return of postmining lands to "approximate original contour" (AOC) to eliminate the highwall has been particularly controversial and technically quite difficult to achieve (Figure 2). A great deal of research and development in mining and reclamation techniques has occurred since the enactment of SMCRA to determine the efficacy of its various requirements and to develop mining and reclamation protocols that will insure that environmental performance standards are met. This chapter details our approaches and findings for environmentally sound mining and reclamation in the steeply sloping lands of the central Appalachian region.

A. The Powell River Project

In 1980, the Penn-Virginia Resources Corporation, a large land holding company, and Virginia Polytechnic Institute and State University (VPI & SU) began a long-term cooperative research/demonstration project in southwest Virginia with the goal of improving the land use potential of surface mined lands in the region (Figure 3). Since that time numerous mining firms, local governments, and state and federal agencies have joined and contributed to the effort which is known as the Powell River Project, named after the initial research watershed (Figure 4). This unique project, combining many different disciplines from the University with the practical approach of industry and the regulatory perspective of various state and federal agencies, has allowed an integrated and objective evaluation of the entire mining process as it relates to postmining land use potential and environmental quality. The results and protocols presented here are the direct result of this cooperative effort.

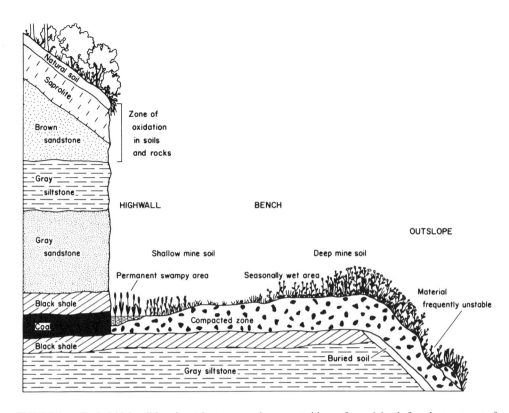

FIGURE 1.. Typical highwall-bench-outslope topography generated by surface mining before the enactment of the 1977 Surface Mining Control and Reclamation Act (SMCRA). Overburden removed from the mining cut was frequently bulldozed over the outslope, generating unstable slope conditions. Sandstones and siltstones dominate the geologic strata in this area, with shallow-lying rocks being oxidized and leached compared to deeper unweathered materials.

FIGURE 2. Surface mined area returned to approximate original contour (AOC) in the steeply sloping central Appalachian region. The resultant landform is steep, erosive, and frequently unstable.

FIGURE 3.   Surface mined land reclaimed to a productive hayland and pasture land-use by controlled overburden placement techniques.

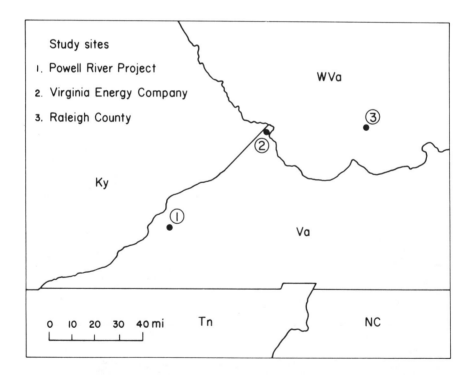

FIGURE 4.   Map of the south-central Appalachian region where this research has been conducted, showing three major study areas. All lie within the Pottsville Group, a coal bearing geologic formation of Pennsylvanian age.

B. Problems in Rehabilitating Surface Mines in the Central Appalachians

We define successful reclamation as a process that prepares mined land for productive postmining use, a use that serves the needs of the landowner and the society at large. Successfully reclaimed mined lands may support vigorous natural growth forests and wildlife populations, agricultural enterprises, or commercial development. Unsuccessfully reclaimed land will not be capable of supporting productive land use. Due to sparse vegetative cover and physical instability, such lands will cause problems for society for many years to come as erosion and water quality hazards develop. If they do not meet the environmental performance standards of the current law (SMCRA), they will cause problems for their owners, as well, who will be required to bear the expense of remedial action.

The natural landscape over much of the central Appalachians is deeply dissected, dominated by steep forested side-slopes and narrow alluvial valleys. The population is concentrated in these valleys due to scarcity of flat land, and flooding is a constant danger. Natural soils over this landscape tend to be thin, rocky, acidic, and infertile. Stripping and stockpiling topsoil before mining is costly and often technically impossible due to steep slopes and the thinness of the soils. Therefore, topsoil substitutes derived from blasted rock spoils must be employed[1] as a plant growth medium during reclamation. Many rock strata are unsuitable for use as topsoil substitutes, however, because of high levels of acid-forming pyrites[2] or other factors that limit productivity. Proper selection, placement, and management of these topsoil substitutes is critical to the establishment and maintenance of the long-term vegetative community required by law.

A second major problem generated by the steep topography is that of disposal and stability of mined spoil. During the mining process, the actual volume of spoil swells by a factor of 1.2 to 1.5 when the hard competent rock is blasted into loose spoil. Thus, even when the highwall is completely eliminated and the mined area is returned to AOC (Figure 2), large volumes of excess spoil usually remain that must be placed into some stable configuration in the landscape. This can be quite challenging in steeply sloped areas, and the material is usually filled into an adjacent hollow or valley. Before the implementation of SMCRA, this mining spoil was frequently bulldozed over the edge of the bench and allowed to cascade downslope (Figure 1). The unstable slopes and landslides generated by this practice were a major factor leading to the inclusion of the AOC provision in SMCRA; presently, all spoil must be controlled on the bench produced by mining or in stable excess spoil disposal fills.

There are certainly many other problems involved in reclaiming steep slope surface mines, but these two (the necessity to control all spoil in a stable fashion and the importance of identifying, isolating, and properly placing the topsoil substitutes) are probably the most difficult, and certainly the most expensive. In the past, these factors were seldom considered as a part of reclamation planning.

Once the proper material is placed at the final reclamation surface, it is essential that an effective strategy be employed to rapidly transform these ''mine spoils'' into ''mine soils'' capable of supporting a vigorous living community over long periods of time. Since most states required only short-term (1 to 2 growing seasons) vegetative communities for bond release before SMCRA, little research has been documented in the U.S. regarding factors critical for the establishment of permanent self-sustaining plant communities on topsoil substitutes.

II. CONTROLLED OVERBURDEN PLACEMENT

A. Use of Natural Topsoils vs. Topsoil Substitutes

Certain geologic strata can be designated for use as topsoil substitutes when it can be shown through premining sampling and testing that their physical and chemical properties surpass those of the native soil present for potential plant growth. However, if sufficient

FIGURE 5.   Active surface mining operation is the central Appalachians. Suitable hard rock spoils produced by the mining process are used as topsoil substitutes on the final reclaimed surface. This requires considerable planning and on-site coordination to insure that the proper strata are isolated, hauled, and spread with sufficient depth at the proper location.

quantities of productive natural soils are present on a site, they should be used as the surface medium whenever possible. While the majority of soils in these Appalachian landscapes are shallow and infertile, occasional deeper soil bodies do occur, particularly in coves and on stable ridgetops. If it is practical to isolate and store sufficient quantities of these materials so that final surface coverage of the site will exceed 0.5 m, there are great advantages to the use of natural soils. The organic matter and microbial populations of natural soils are invaluable to the revegetation process.

Regardless of the quality of the natural soil, however, it must be present in large enough quantities to allow sufficient thickness on the final reclamation surface. Quite frequently, only 15 to 40 cm of natural soil are spread over highly compacted spoil materials, leading to rooting restrictions and problems of seasonal wetness. It is also important that the true topsoil (A plus E horizons) be separated from the underlying soil and rock horizons during stripping, and then be stored properly to maintain the viability of aerobic microbial populations.[3] Due to the thin nature of most of the surface soil horizons in this region, when natural topsoils are employed in reclamation they are usually a mixture of all soil horizons above hard bedrock. Thus, the positive influences of their organic matter and microbial content are considerably reduced. Many mining operators do find it beneficial, however, to mix whatever true topsoil becomes available with their designated topsoil substitutes during final reclamation grading.

B. Selecting Geologic Materials for Use as Topsoil Substitutes

The rock beds in the southwest Virginia coal fields lie essentially flat with a gentle dip to the northwest. The coal seams are separated by varying thicknesses of fluvial-deltaic sandstones and siltstones, with a minor component of shales. Commonly, multiple seams of coal are mined, with all strata above the lower seam being blasted into spoils and handled in some fashion during the mining operation (Figure 5). In order for these hard rock spoils to be successfully employed as topsoil substitutes, the optimal strata must be identified before mining so that the mining plan can be designed to place the proper strata at the final reclaimed surface. Also, any potentially toxic strata must be identified and then isolated away from the surface and local groundwater.

Before mining, the thickness and variability of the various overburden strata are determined by exploratory drilling. The exact spacing of bore-holes required to accurately characterize a given strata varies tremendously, and the experience of the operator, permitting personnel, and regulatory agencies must be relied upon to insure that representative samples are used for testing. All major strata are tested unless previous experience indicates that one particular stratum is superior.

In general, the following criteria, in order of importance, are essential to evaluate the suitability of a given strata for use as a topsoil substitute: (1) acid-base accounting, (2) rock type, (3) extractable nutrients, (4) pH, (5) soluble salts, (6) degree of weathering and oxidation before mining. Specifics of interpreting the various tests for this region are given by a number of authors.[4-6]

Ideally, a stratum or multiple strata are isolated, which will be nonacid forming over time, and therefore high in pH and low in soluble salts. Rocks that are low in acid-forming pyrites (FeS_2) and high in carbonate cementing agents are ideal. Strata with net acidities in excess of 2 ppt (or ~2 tons of lime/acre) should be regarded as potentially toxic and isolated during mining. Rock type is important since spoils derived entirely from sandstones tend to be very coarse and droughty, while those derived entirely from fine siltstones and shales tend to form hard surface crusts and impede water percolation. Several researchers[7-9] have found that mixtures of rock types are superior to those composed of all siltstone or sandstone. Extractable N and P are generally low in all spoils,[10] but strata with significant carbonate cementation will supply large amounts of plant available Ca and Mg over time. K availability is seldom a problem in these materials,[11] as long as acid generating spoils are avoided.

Spoil pH per se is an erratic indicator of spoil quality since it does not take into account the acid-forming potential of unweathered pyrites. In weathered, near-surface strata, however, materials with pH below 5 should be avoided. Quite often, these leached and oxidized strata are preferred by the mining operators for use as topsoil substitutes because they blast into a finer, less rocky spoil that is easily handled and spread. These brownish-red, oxidized materials are usually high in Fe-oxides, however, which can be detrimental to long-term P availability due to their capacity to permanently adsorb or "fix" applied P-fertilizers.[7] Another advantage to using the finer, preweathered strata is that they will hold more plant-available water due to their lower rock content. The majority of unweathered strata blast into spoils that contain 30 to 50% soil sized (<2 mm) fragments and will supply sufficient plant-available water as long as they are placed at the final surface with sufficient uncompacted depth.

Perhaps the most important criteria for selecting a topsoil substitute is whether or not the designated strata can be isolated and handled within the mining plan without excessive cost to the operator. If an ideal stratum is thinner than the usual blasting lift thickness or placed inopportunely within the geologic column, its use may be impossible. Quite often, two or more adjacent strata within the same blasting lift will be identified as the substitute materials, and then handled and spread together. In this fashion, strata with dissimilar physical and chemical properties can be mixed together into a composite with properties more favorable than those of individual strata.

C. Mine Soil Construction to Insure Reclamation Success

Before the enactment of SMCRA, little thought was given to selective handling of overburden. The final surface generated for revegetation was usually a rough graded mixture of all strata present in the overburden column, leading to extreme heterogeneity in mine soil

FIGURE 6.   Final reclamation grading of topsoil substitute mine spoils. A minimum of 1 m of nonacidic spoil of known geologic origin has been carefully placed on this final reclamation surface. Vehicle traffic has been excluded from the area to prevent excessive compaction, and final grading is performed with light equipment to further minimize compaction.

properties on these older benches. It is not uncommon in these areas to find pH 3 and pH 8 mine soils directly adjacent to one another. This spoil variability, combined with problems of severe compaction, makes it difficult to develop uniform management strategies for many of these older benches. The primary objective of modern controlled overburden placement techniques is to place a designated topsoil substitute of controlled geologic origin at the final reclamation surface. This final lift of spoil must be placed with sufficient thickness (>1 m) to support vigorous plant growth over time, and must not be excessively compacted.

The importance of avoiding compaction cannot be overemphasized. In a study of older mine soils (5 to 20 years) in the Powell River Project watershed, Daniels and Amos[12] found that compaction was the major soil factor limiting long-term revegetation success. Severely compacted (bulk density >1.7 g/cc) mine soils, particularly those with <50 cm of effective rooting depth simply cannot hold enough plant-available water to sustain vigorous plant communities through protracted drought. Conversely, these compacted zones may also perch water tables in unexpected locations, causing saturation and undesirable anaerobic conditions within the rooting zone. These compacted zones result from the repeated traffic of rubber tired loaders and haulers, and bulldozers to a lesser extent.

Continuous on-site coordination and supervision by the mining operator or job foreman is necessary to insure that the designated strata are correctly isolated and hauled to the final reclamation surface. First, the reclamation area is filled with nonselected spoil to a level 1 to 1.5 m below the planned final surface, and rough graded. Then, the entire area is dumped with the appropriate spoil in closely spaced piles. The spoil may remain in this configuration indefinately before the final reclamation grading is performed in one operation (Figure 6). Grading should be delayed until just before seeding whenever possible to prevent surface crusts from forming. This will also minimize surface runoff and erosion. Grading wet spoils promotes compaction and should be avoided whenever possible. By following these procedures, thick, uniform, uncompacted mine soils can be produced with few direct costs to the operator (other than those involved with coordination and supervision) since all of the materials must be handled and moved regardless of placement location. Throughout the process it is important to maintain alternate spoil dumpsites so that acidic or extremely coarse spoils are eliminated from the final surface. The key factor is control of overburden handling and movement for the sake of improved reclamation success.

III. ESTABLISHING AND MANAGING VEGETATION

A. Revegetation Strategies and SMCRA

Assuming that a suitable plant growth medium has been generated for revegetation, the choice of species depends primarily on the intended postmining land use. The majority of lands in the central Appalachians are intended to return to native forest land use, and therefore the primary goal of revegetation are erosion control and soil building to allow for productive forest growth over time. For unmanaged, postmining lands uses, such as natural forest growth, SMCRA requires that a vigorous self-sustaining plant community persist on the site for 5 full years after the last application of lime, fertilizer, or seed. At the end of the 5-year period, at least 90% living ground cover must be maintained.

In order to meet the 5-year bond release goal, the mix of species used must be capable of rapid uptake of large amounts of applied fertilizers during initial growing seasons; and, in subsequent growing seasons, these nutrients, particularly P, must be cycled to meet plant needs. N-fixing legume species must establish and then persist for many years to build the soil-N pool and insure long-term N availability. Perhaps most importantly, soil microbial populations critical to the decomposition of nutrient-containing plant residues and for the mineralization of organically complexed N and P must become established. For all of this to occur within the soil/plant community over a 5-year period without augmentation requires a carefully planned revegetation strategy, and a little luck. The augmentation restriction is waived for higher land uses such as hayland and pasture or intensive forestry where periodic management or augmentation is a normal management practice, but few lands returned to approximate original contour in the central Appalachians are really suitable for these uses.

B. Initial Establishment of Vegetation

Assuming that a suitable growing medium has been produced as a result of controlled overburden placement, the most important factor controlling the successful establishment of vegetation is the time of seeding. Early spring and fall plantings have the highest chances of success, while mid-summer plantings are very risky. In general, slower growing perennial grasses and legumes are sown along with a fast growing annual cover or nurse crop. The rapid establishment of a cover crop serves many functions. It must provide rapid initial ground cover for erosion control and partial shade and mulch for the germinating perennials. Perhaps most importantly, the cover crop takes up applied N and P fertilizer, concentrates it into organic forms, and then releases it upon decomposition of first year litter. This is most important for N retention, since large amounts of applied fertilizer-N can be leached away if heavy rains occur before the plant community has established extensive root systems.[13]

Various hydro-seeding techniques are used to deliver seed, fertilizer, lime, and mulching agents simultaneously. Typical liming, seeding, and fertilization recommendations for this region are given by McCart and Daniels.[14] It is particularly important that legume species capable of persisting for long periods with low soil-P levels be selected. Agronomic species developed for high fertility environments generally decline rapidly after the first several growing seasons. Extra mulching materials greatly enhance success, particularly on harsh sites.[15] Another advantage to using natural organic mulches like straw or shredded bark is

Table 1
**CHANGES IN STANDING BIOMASS (*FESTUCA ARUNDINACEAE*
SCHREB.) IN RESPONSE TO VARIOUS SURFACE TREATMENTS
IN OCTOBER OF 1982 THROUGH 1984, AND SOIL ORGANIC
MATTER AND TOTAL-N CONTENT IN 1984[a]**

Treatment	Standing biomass (Mg/ha)			Organic matter 1984 (%)	Total-N 1984 (%)
	1982	1983	1984		
Control (NPK only)	5.69b	6.03c	0.94d	2.24e	0.097d
Topsoil (30 cm + NPK)	5.80b	5.16c	1.71cd	1.36f	0.059d
Sawdust (112 mg/ha + NPK)	3.67c	5.56c	2.76c	8.47a	0.198bc
Sludge (22 mg/ha)	3.92c	5.20c	2.74c	3.66d	0.171c
Sludge (56 mg/ha)	7.66a	10.08b	4.44b	4.81c	0.251b
Sludge (112 mg/ha)	9.27a	13.57ab	4.84b	6.99b	0.457a
Sludge (224 mg/ha)	9.01a	16.48a	7.49a	6.42b	0.419a

Note: Means within columns followed by different letters are significnatly different ($\alpha = 0.05$).

[a]   These data are dervied from a controlled overburden placement experiment,[21] and all treatments
were seeded into a topsoil substitute derived from sandstone and siltstone spoils. Note the
sharp declines in all treatments in 1984, and in particular the drastic decline in those treatments
not receiving organic amendments. These plots did not contain legumes, and therefore N
stress became acute in 1984 once initial fertilizer effects passed. Municipal sewage sludge is
a superior surface amendment.

that they will speed the introduction of microbes into the system. Legume seeds are inno-
culated with their specific *Rhizobium* symbionts before seeding, and care must be taken to
insure that the strong fertilizer solutions used in seeding are not toxic to these essential
microbes.[16]

Tree seedlings are also planted during the first growing season when specified by the
postmining land use. However, the rapid initial growth of the herbaceous vegetation used
to control erosion and stabilize the site competes heavily with the young trees.[17] This is a
particular problem when the postmining land use is specified as return to native forest, since
most state regulations require 90% ground cover and establishment of several pioneer tree
species simultaneously. Use of less competitive herbaceous vegetation or local weed control
will enhance forest land use success.

C. Factors Critical to Long-Term Revegetation Success

While vegetation establishment is an important first step, the law requires that a vigorous
plant community persist for at least 5 years. This can be exceedingly difficult when augmentation is not allowed. Initial fertilization effects will usually last for the first two growing
seasons, assuming good initial establishment. After that time, steady decreases in standing
biomass and ground cover are common, even on the best of mine soils (Table 1). Assuming
adequate initial mine soil conditions, the long-term productivity of the plant/soil system is
dependent on two major factors: (1) the accumulation of organic matter and N, and (2) the
establishment of an organic-P pool and the avoidance of P-fixation. Both of these are in
turn highly dependent on the introduction and function of microbial communities over time.

Mine spoils are essentially devoid of N initially, so the total amount of N required to
sustain plant growth over time must come from initial fertilization and subsequent symbiotic
N-fixation by legumes. Working in England, Dancer et al.[18] found that approximately 700
kg/ha of ecosystem N were required before native woody species could invade stabilized
china clay wastes. Usually <150 kg/ha of N are added as fertilizer, and much of this may

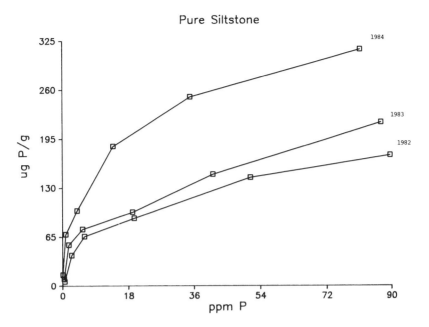

FIGURE 7. P adsorption isotherms for a southwest Virginia mine soil over three growing seasons. The isotherms indicate increasing P adsorption onto soil (μg P/g) as soil solution P (ppm P) increases. The consistent increase in P adsorption capacity with time indicates that over the 5-year revegetation bond release period mandated by SMCRA, P availability is likely to limit plant growth. This increase in "P-fixing" potential with time is due to the weathering and oxidation of ion compounds in the original spoil. The resultant iron oxides have a high specific adsorption affinity for P in solid solution, and over time the adsorbed P is converted into unavailable solid forms.

be subject to leaching losses. The vast majority of N needed to supply plant/soil community needs must, therefore, come from symbiotic N-fixation and subsequent mineralization of organically combined N. While small amounts of N may also be fixed by free-living microbes,[19] it is obvious that the maintenance of a vigorous legume component within the plant community is critical. Many of the agronomic species commonly employed are either short-lived or biennial and are generally not adequate. Species like black locust (*Robinia pseudoaccacia*) and *Sericea lespedeza* have been used quite successfully for many years on a wide range of spoil types,[20] but their use has dropped off in recent years due to increased use of more traditional forage legumes and grasses in reclamation mixtures.

Since N is primarily combined in organic matter in soils, the addition of organic amendments to the soil can greatly enhance total soil N and N availability over time (Table 1). Sewage sludge has been shown to be a particularly effective mine soil amendment in numerous[21,22] studies, but may not always be available in sufficient quantities for use on remote sites. Local and state regulations frequently complicate the use of sewage wastes on disturbed lands. Sawdust and bark mulch are also helpful in increasing the initial mine soil organic matter content, but are generally low in N content.

The maintenance of plant available P over time is hindered by two factors: (1) fresh geologic materials are generally low in readily plant available P and (2) as mine soils weather and oxidize they become enriched in Fe-oxides that specifically adsorb P from solution and then "fix" P into unavailable forms. This propensity of mine soils to fix P increases over time (Figure 7). It is therefore critical to establish and build an organic-P reservoir (Figure 8) in the soil to supply long-term plant needs through P mineralization. Large fertilizer applications of P (>250 kg/ha) are required to insure that sufficient labile P will be available

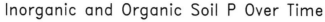

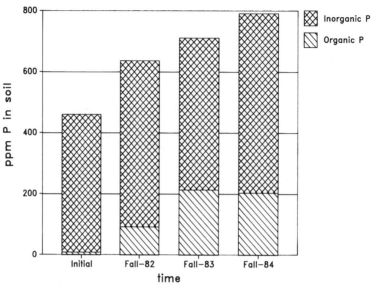

FIGURE 8.   Changes in the form of total P in the surface of a southwest Virginia mine
soil with time. Total P increases with time as the plant community concentrates available
P into the surface soil. Due to the high P-fixing potential of these soils, the development
of an organic-P reservoir over time and subsequent mineralization of this fraction is
critical to long-term P availability.

over the first several years to supply plant growth and build the organic-P pool. Smaller
amounts of P will also become available to the plant community as native calcium phosphates
in the rocks decompose,[9] but this P is not sufficient to meet the needs of a vigorous plant
community. Many plant species, particularly those which are mycorrhizal, are able to draw
P from difficultly available sources and should be used whenever possible.

Over the 5-year bond release period, it is likely that N will first limit plant growth due
to greater plant needs, and that P fixation will become a problem in later seasons as the
mine soils weather and oxidize. However, if plant production is limited by N, the transfer
of fertilizer P into organic forms will be limited, thus increasing P fixation. Similarly, low
soil P levels may also hinder N accumulation since symbiotic N-fixing bacteria have a high
P demand.[23]

IV. USING MINING SPOIL TO CONSTRUCT POSTMINING LANDFORMS

The second major problem generated by surface mining in steep topography is spoil
disposal. The predominant spoil disposal practice for surface mining firms in the Appalachian
region is to conform with the letter of the law, to dispose of the majority of mined spoil in
the contour mining "cut" upon the mined bench. In steep slope contour mining situations,
additional spoil disposal areas must also be prepared, filled, and reclaimed due to the
"swelling" of rock when it is blasted and the necessity to dispose of the excess. These
excess spoil disposal fills may be constructed in a variety of locations, including hollows
and abandoned pre-SMCRA mining benches, with a primary constraint being proximity to
the mining area due to the cost of transport and the large excess spoil volumes. The primary
legal requirements for these excess spoil disposal fills is that they be constructed in stable
fashion and revegetated.

There are two situations where the law allows exceptions to AOC provisions. If the entire top of a mountain lies above a coal seam that is to be totally removed, the mining will be considered as "mountaintop removal — valley fill" mining, and the mining firm may leave the mountaintop as a broad, near level area. However, in the practice of mountaintop removal mining in steeply sloping regions, excess spoil disposal becomes a problem and operators often find spoil disposal best accomplished by placing the maximum amount of spoil upon the mined bench, and thus reconstructing a facsimile of the original mountaintop. The other exception occurs where the mining operator and the landowner attest to their intention to prepare the mined land to a more intensive land use than that which preceded mining. In this case, the law requires that documentation be provided with the permit application to support the postmining land use intentions. In practice, this also seldom occurs. Thus, the majority of mined lands in the steeply sloping regions are returned to AOC at the conclusion of mining activities.

Reclamation to AOC is of recent origin in Virginia and neighboring states where pre-SMCRA regulations did not require the backfilling of highwalls. As a result, some of the problems with this practice are just coming to light. These problems include difficulties in the use of controlled overburden placement procedures, unstable highwall backfills, and foregone opportunities to construct more favorable landforms.

A. Application of Controlled Overburden Placement Procedures on Lands Returned to AOC

In order for a mining operator to devote the attention to surface media construction required by controlled overburden placement procedures, he must have some incentive to do so. The land use potential of steeply sloping AOC topography will not, in most cases, provide that incentive. Steeply sloping unmanaged hardwood forests are abundant in the central Appalachians, and their land use value is low. Therefore, mining operators have a difficult time seeing any practical reason for going beyond the minimum procedures required for bond release. Furthermore, the perceptions of many operators regarding reclamation activities required for bond release may not be correct, since their primary experience has been with a 2-year bond release period rather than the currently mandated 5 years.

In addition, the cost-effective use of controlled overburden placement procedures requires that multiple spoil disposal areas be available relatively close to the mining pit. Generally AOC mining does not meet this condition (Figure 9). Contour mining is generally a linear process, with primary spoil disposal activities occurring in the mining pits recently vacated by active mining. Within these pits, AOC backfills are generally constructed in discrete sections. Each section is constructed from the bottom up, and the area required by a section of highwall backfill increases with highwall height due to hauler ramping requirements. As backfill construction proceeds, previously placed spoil tends to be covered by materials cascading down from above. The result is that it is difficult to maintain an active highwall backfill close to the mining pit in condition to receive delivery of suitable surface materials as they become available. Yet, the cost-effective use of controlled overburden placement requires that the designated strata be loaded into trucks and removed as they are encountered in the course of mining and delivered directly to the final reclamation surface. Generally speaking, the spatial limitations of AOC highwall backfill construction in steeply sloping areas do not provide mining operators with sufficient spoil handling flexibility to use controlled overburden placement procedures cost effectively.

B. AOC Backfill Stability

An additional problem with AOC backfills is the potential for instability in situations where backfills are improperly constructed, subjected to excessive wetness, or are excessively steep. Bell and Daniels[24] studied a number of AOC sites where slope failure had occurred,

FIGURE 9.   A typical AOC backfill under construction. Common construction procedures include extension of hauling route ramps as the dumping point gets closer to the top of the highwall.

or appeared imminent, and performed detailed stability analyses and modeling on four of them. Improper construction techniques were observed to include placement of excess spoil below the mined bench, in direct violation of the current law where slopes exceed 20°. However, it was also observed that these mining situations presented the mining firms with no economical alternative, due to lack of adjacent areas suitable for excess spoil disposal.

Seepage of water into the spoil was also determined to be a factor that contributes to backfilled spoil instability. Since coal seams often act as aquifers in the Appalachians, seepage is a frequent problem. Where seepage is anticipated, it can be compensated for by fill design. However, seeps generally are not mapped, and their location may not be known until they are encountered in the course of mining. Seasonal seeps may not be recognized during mining if they are exposed during the dry season. If unanticipated seeps are encountered, it is possible to develop strategies for dealing with them on site. However, this may entail a stoppage or slowdown of production and definitely requires close coordination between engineering personnel and the jobsite. Thus, there are many instances where seepage control is not built into the backfill. The results may include slope instability due to buildup of pore water pressure, and/or reduction of spoil shear strength due to rapid moisture-induced weathering of the rocks and minerals composing the spoil material.

Excessive steepness also contributes to highwall backfill instability. Natural slopes in the Appalachians sometimes exceed 30°. The effects of steepness are aggravated in situations where wetness is a problem. Another contributing factor is the difficulty of achieving uniform compaction in a steeply sloping highwall backfill, in spite of common engineering design assumptions of uniform compaction. Bell and Daniels found evidence of instability (factors of safety below 1.0), in two backfills that showed no evidence of wetness. These two backfills were, in certain respects, typical of many backfills in the region. It is ironic that the supposed "solution" to the pre-SMCRA problem of unregulated and unstable over-the-outslope spoil disposal may prove to be the cause of similar problems in future years.

C. An Alternative to AOC

Consideration of the final form of the reclaimed land, in the context of the region as a whole, compounds the problem of spoil disposal. The Appalachian region is subject to a

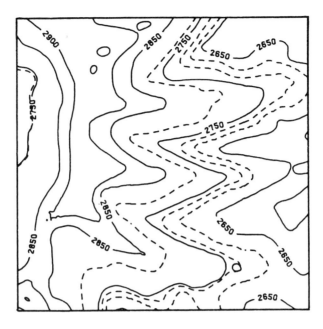

FIGURE 10. Premining contours at Amos Ridge. The area represented is approximately 40 ha. Dotted lines represent coal outcrops. The first hollow fill was constructed at a location centered approximately 200 m north of the southern map border, below the middle coal seam.

number of land use problems as a direct result of its steeply sloping topography. Few who are familiar with the region would argue with the contention that a lack of flat land has severely limited economic development. Level, nonflood-prone housing sites are scarce and expensive, agricultural and industrial sites are nearly nonexistent. As a result, the coal industry is the major employer of the region. In times when the price of coal is low, local unemployment rates are high in spite of general prosperity. This situation persists in spite of the fact that the coal industry possesses the physical means to transform at least portions of the landscape, so as to partially alleviate the developmental constraints imposed by the scarcity of flat land. However, for a variety of reasons, that potential is not being realized at present.

An alternative to AOC in steeply sloping topography has been investigated at a surface mine at Amos Ridge, in Wise County, Va. At this site, the premining topography consisted of a series of finger ridges protruding from a central "spine", Amos Ridge (Figures 10 and 11). Excepting the tops of the fingers, nearly all the land being mined has had slopes in excess of 20°. This type of topography is common throughout the region.

This site is being mined under an experimental variance from the provisions of SMCRA, obtained with the cooperation of the U.S. Office of Surface Mining and the Virginia Division of Mined Land Reclamation. Three hollow fills are being constructed so that their upper surfaces are contiguous with flat areas at the tops of the finger ridges, which are not being returned to their original heights (Figure 12a). The result is that a relatively large near-level area (approximately 5 of the 29 ha under permit) is being constructed over the stripped fingers and filled hollows. The objective of producing usable land is being further pursued by constructing the hollow fill outslopes at 3:1 grades, rather than the conventional 2:1, and by using 1 m of uncompacted soil and spoil materials to construct productive topsoil substitutes on the near-level upper surfaces. With the exception of the AOC and hollow fill

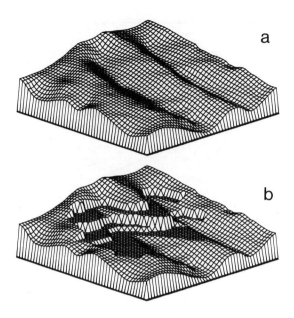

FIGURE 11.   Isometric representations of premining and mined landforms at Amos Ridge. The area represented is ~20 ha, and includes the southeastern corner of Figure 9. (a) The topography previous to mining, showing finger protruding from the central "spine" of Amos Ridge at the western edge of the image. (b) A representation of the portion of the topography disturbed by mining during the period of study. The site never took this exact appearance, since the economics of spoil handling dictate minimization of spoil movement distance.

construction provisions (two of the three hollow fills are being constructed using experimental techniques), all environmental performance standards of SMCRA are being met.

The mining and reclamation operations at this site are being studied and monitored intensively. The hollow fills are being monitored for stability and water quality. In addition, the costs of producing this landscape have been documented.[25,26] Data collected over the period between January 1, 1984 and August 1, 1985 indicate that the cost of producing this landscape was not excessive (Table 2). During this period, the first hollow fill (HF1) was constructed and reclaimed, the second hollow fill (HF2) was partially completed, the stripped finger point immediately south of the first hollow fill was completely reclaimed (NLA), and highwalls were backfilled above (west of) the finger point and first hollow fill (HWBF) and in areas south of the experimental practice permit area which had been mined in 1983 (AOC). The $1.90 per bank cubic yard overall cost figure compares favorably with informal estimates privately provided by industry personnel for mining operations using more conventional spoil handling practices. The direct cost of reclamation (Table 3) composes a relatively minor portion of the overall cost.

However, looking at the reclamation cost more closely shows distinct differences between the various reclamation areas (Table 4). The cost of isolating favorable surface materials, as required by controlled overburden placement procedures, was found to be a relatively minor cost component. During the period under study, topsoil was the primary material used for surface reconstruction. Due to the necessity to dispose of topsoil covering mineable coal, the only recognized cost of topsoil segregation is the increase in haulage time required to bring the topsoil to reclamation areas rather than to routine disposal sites. The daily records of hauler operation allow estimation of this cost difference. The cost of topsoil segregation

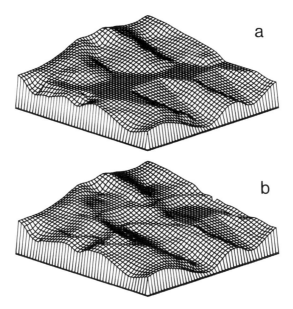

FIGURE 12. Isometric representation of the postmining land-form alternatives at Amos Ridge. (a) A landform alternative (LA) currently under construction. Hollow fill faces are represented by the triangular-shaped surfaces located between the finger ridges at the lower right hand side of the image. (b) A possible landform, had the site been reclaimed to AOC. An access road is located approximately 15 m above the outcrop of the lower coal seam. Excess spoil is disposed in the first hollow above the mining bench and in the fill constructed in the second hollow.

was found to be greatest for the areas being reclaimed to near-level topography, primarily because more topsoil was segregated for use in these areas than on the highwall backfills. As discussed previously, the operational difficulties inherent in reclaiming highwall backfills and lack of operator incentives hinder attempts to isolate favorable material for use in these areas over prolonged periods. However, the difference in grading costs between the highwall backfill and the near-level finger point area was more than sufficient to compensate for differences in topsoil isolation expense. The high cost of grading the hollow fill surface was due to the frequency of water drainage structures. One can surmise that the expense of isolating a particular rock layer may be less than that of isolating topsoil, since topsoil is a relatively scarce material in most Appalachian landscapes. Of course, the above costs do not include the expense of personnel time required to establish coordination between reclamation and spoil placement activities. This will vary widely from site to site.

D. Comparative Economics of Landform Alternatives

The last section considers reclamation costs as a process of preparing the surface of a reclaimed landscape, given that landscape. However, in the context of the land use problems of central Appalachian communities, we believe it is more appropriate to consider the process of reconstructing a mined landscape as the reclamation activity. As a part of our studies at the Amos Ridge mining site, we have been developing computer based methods for estimating the relative costs of reconstructing alternative postmining landforms.

The landform construction cost estimating system has three primary components. COSTSUM[27] is a set of seven FORTRAN programs designed to analyze operational cost

Table 2
COMPONENTS OF
AVERAGE SPOIL
HANDLING COST AT
AMOS RIDGE

Component	$ per bcy[a]
Clear and bench	0.01
Drill and blast	0.41
Carry and push	0.20
Load and haul	0.89
Reclamation[b]	0.08
Overhead	0.31
Total	1.90

[a]   A bank cubic yard (bcy) is a measure of spoil volume before disturbance and swell.
[b]   Direct costs listed in Table 3.

Table 3
OPERATIONAL COSTS CONSIDERED AS
RECLAMATION EXPENSES AT AMOS RIDGE

Direct	Indirect
Area preparation[a]	Haul spoil[b]
Shape dumped material	Work haul roads and dumpsite[b]
Revegetation	
Haul topsoil[b]	
Grading	
Fertilization and seeding	

[a]   Hollow fill construction only.
[b]   Calculated as difference in cost between disposal in specified area and routine disposal.

data from active surface mining sites. The use of these programs provides detailed estimates of spoil handling and reclamation costs. It will also calculate values for many of the input parameters of TOPSIM,[28] a surface mining simulator. TOPSIM is designed to model haulback mining, as commonly practiced in Appalachian regions. Its primary use is in the comparison of cost effects of alternative overburden handling (i.e., postmining landform construction) plans. The three primary modes of spoil movement modeled by TOPSIM are pushing with a dozer, carrying with a loader, and hauling by truck. The third component is CPS/PC, a software system for gridding, contouring, mapping, and analysis of three dimensional data developed by Radian Corporation.[29] The primary uses of CPS/PC are to develop volume estimates of the stripping blocks, and of the fill cells composing postmining topographies under consideration, and to prepare graphic images of the postmining landforms as a part of the planning process (Figure 12).

We used this cost-estimating system to model the economics of landform construction at Amos Ridge, by comparing the actual postmining landform (the landform alteration, or LA, case; Figure 12a) to a likely alternative, had the site been mined and reclaimed using conventional methods (the AOC case; Figure 12b). Excess spoil disposal for the AOC case

Table 4
RECLAMATION COST COMPONENTS AT AMOS
RIDGE, PER ACRE AND PER BCY, BY RECLAMATION
AREA[a]

Cost	HF1 ($)	NLA ($)	AOC ($)	HWBF ($)
Per acre				
Haul topsoil	817	451	337	172
Finish grade	1770	537	1412	1671
Fertilize and seed	770	965	750	1233
Total revegetation	3376	1950	2500	3076
Per bcy				
Revegetation	0.07	0.04	0.04	0.05
Total direct	0.19	0.04	0.04	0.05
Total: direct plus indirect[b]	0.19	0.04	0.09	0.06

[a]   See text for explanation of symbols.
[b]   Indirect cost does not include higher than average hauling costs from stripping
       blocks primarily disposed in AOC area.

is accomplished with a single hollow fill in the second hollow, so as to minimize permitting and bonding costs. Additional excess spoil is disposed in the first hollow above the bench produced by mining the lower coal seam, in conformance with observed practices. The postmining landform on the southernmost finger ridge contains slopes reconstructed to their original 2:1 slope, with an access road located approximately 40 ft above the outcrop of the lower seam.

The results of modeling procedures indicate that the cost of constructing the LA case was actually less than what the cost would have been had the AOC alternative been pursued.[30] Depending upon the spoil movement assumptions utilized, the estimated cost difference varies between $0.14 and $0.58/ton of coal produced. Given the current $15.00 to $20.00/ ton selling price, this cost difference is small but significant. The primary difference between the two landforms is the cost of hauling. To construct the AOC landscape, a greater quantity of material must be hauled; since the first hollow is not being filled, materials that were pushed or carried directly into the first hollow fill in the LA case required hauling in the AOC case. Also, longer, steeper haul distances are involved in the AOC case, since much of the material that was moved downward from its point of origin into the first hollow fill in the LA case must be hauled upwards, to rebuild the finger ridge at the southern edge of the site. Haulage is generally a more expensive mode of material movement than direct dozer push or loader carry, and, obviously, longer, steeper upward hauls will be more expensive than the short downhill hauls into the first hollow fill. To a certain extent, this difference in haulage cost is compensated, since the cost of constructing the first hollow fill need not be borne if constructing the AOC topography. Another cost advantage held by the AOC case, as modeled, is the absence of increased haulage costs required to isolate topsoil or other surface media, due to the impracticality of preparing steep slopes for an improved land use and the generally favorable qualities of the overburden at the site. However, excluding the hollow fill, per acre reclamation costs would have been greater in the AOC case due to the steeply sloping surfaces (Table 4). The net cost balance appears to favor the altered landform (Figure 12a).

To complete discussion of the comparative economics of alternative landforms, factors that were not modeled must be considered. From the point of view of the operator, the major nonquantified advantage to the AOC case would be less engagement with unfamiliar hollow fill construction techniques. However, this advantage is offset by disadvantages. The LA case offers the mining operator increased operational flexibility. The near-level areas are used for equipment storage and maintenance during the course of mining. Also, due to the opportunity to haul, carry, and push laterally into the first hollow fill, it gives greater opportunity to avoid steep uphill hauls on bad weather days, when slick roads reduce hauler efficiency. Also, the LA case will require less wear and tear on his machinery due to the reduction in steep uphill hauling and steep slope grading. Finally, there is a safety factor to consider due to the inherent danger of operating large machines at the top of steeply sloping unconsolidated spoil banks, a situation that occurs less frequently during construction of the LA landform.

From the point of view of the landowner, there are also advantages to the LA case. It would seem that the primary advantage will be improved land use potential. In the immediate future, the near-level areas and hollow fill faces at Amos Ridge will be used as pasture and hayland. We expect its soils to be productive, relative to other reclaimed and natural pastured areas. In the long term, a variety of other uses will be possible due to favorable topography. Although it is doubtful that this particular site would be remined using surface methods, the ease of remining a site reclaimed to a topography similar to the LA case, relative to the AOC case where far more spoil would have to be moved in order to expose the coal seam, would be a definite advantage to a landowner with a continued interest in mineral exploitation and substantial remaining mineral deposits.

From the standpoint of society, there also appear to be advantages to the LA case. Primary among these would be a less adverse environmental impact. The potential for construction of unstable highwall backfills is minimized through use of the LA technique, since highwall backfills are themselves minimized and the resultant topography gives the mining firm spoil handling flexibility. In the event of a failure, the existence of the near level area constructed below the highwall would localize the adverse impacts. It also seems that the LA case would have more favorable hydrologic consequences. The steep reconstructed slopes of the AOC landform are interrupted only by the access road and narrow terraces. In contrast, the broad terrace of the LA landform will have the effect of dispersing, rather than concentrating, water moving downslope from the terrain above, and it would be far more effective at reducing reclaimed landscape sediment yield than the AOC landform.[31] In a broader sense, the LA landform will also have the effect of redistributing downward moving water in time, since some would be temporarily stored in the soils and spoils on the near-level area, reducing storm related peak flows that cause flooding in steeply sloping areas.

V. INSTITUTIONAL CONSTRAINTS TO IMPROVED RECLAMATION

In spite of opportunities to perform improved reclamation, the vast amount of land being mined in the region is being returned to AOC. Thus, there is potential for landscape improvement through coal surface mining that is not being realized at present. When we ask why, we do not see technical knowledge as the limiting factor. To the contrary, technologies have been demonstrated that have the potential to improve reconstructed mine soils over their native counterparts, in terms of productivity. The surface mining industry has long possessed the power to transform the landscape to alleviate the chronic shortage of nonflood-prone, flat lands that constrains the economic development of the central Appalachian region. It appears that the cost of producing deep productive soils upon near level areas in characteristic steeply sloping points-and-hollows topography will be less, in some instances, than the cost of the common practice of reconstructing steep original slopes.

As with most things, it is an imperfect situation, and there are many factors interacting in complex fashion to cause it. However, in this case, the primary factors appear to be institutional, a product of the variety of histories, goals, and constraints that influence the behavior of mining firms, land owners, and regulatory personnel. Certainly, it is easy to point the finger at the current law, particularly its AOC provisions. However, the current law is what forms the present context. We contend that situations exist where it is in the interest of the mining firms to produce stable, productive landscapes, given the commonly practiced AOC option as an alternative. However, in the absence of law, it was in the interest of individual mining firms to push the majority of mined spoil over the outslope. Development of effective laws and regulations is a learning process, and it is only with experience of the effects of SMCRA that possibilities for improvement have become apparent.

A primary effect of SMCRA has been to require the production of AOC backfills in steeply sloping regions. As discussed above, there are situations in which such constructions might not be the most beneficial reclamation practice. In addition, the rigidity of the AOC provisions result in a situation where, because of disparity in cost between placing spoil in the highwall backfill vs. an excess spoil disposal fill, the mining operator is faced with a financial incentive to deviate from the backfill construction technique that would produce the maximum factor of safety. As noted by Bell and Daniels,[24] if excess spoil disposal is expensive, relative to placement in the backfill, the operator may find it cost effective to place spoil below the toe of the fill. Another means of increasing the amount of spoil in the backfill is to produce a "pregnant" backfill, one that bulges in midslope thereby producing steep lower slopes and possible instability. The converse occurs where backfill placement is expensive, relative to the excess spoil disposal. The law is clear in its requirement that the highwall be completely covered with spoil over the majority of its length. Thus, the tendency is to produce a concave surface with a steeply sloping upper slope. The majority of AOC backfills do not suffer in any of these faults in any obvious way. However, in spite of the legal requirement that AOC backfills be designed to a factor of safety of at least 1.3, there are situations that cause the operator at the site to deviate from engineering plans.

The law does provide a means for obtaining a variance from the AOC provisions in cases where a higher land use is planned than that which precedes mining. However, this provision is seldom taken advantage of. In order to obtain such a variance, documentation must be submitted with the initial permit application, since the permit defines the postmining land use. Depending upon the land use, such documentation may be quite costly. The applicant is required to submit materials substantiating stated intentions and an ability to carry out that proposed land use, as well as design specifications. The cost of producing this documentation must be borne previous to the permit application. The permit application generally precedes the onset of mining by 6 months to 2 years, and the mining process itself may take a few years to complete. There is often an additional time lag between reclamation and land use implementation; thus, many years separate the initial application from the financial returns of the improved land use. Given the time value of money, and assuming that the land owner/permit applicant expects the profits from the postmining land use to pay for the permitting and design costs, this time lag, in effect, handicaps the potential for profitability. The permittee faces another penalty, if the procedure required to prepare the land for its postmining use is a more expensive procedure than that which is commonly practiced, since the amount of performance bond must be sufficient to cover reclamation to the permittees declared postmining use.

The uncertainty factor also tends to inhibit improved use permit applications. Unforeseen fluctuations in the price of coal often cause mining firms to abandon mining plans after they have been permitted but before mining begins. Similarly, unforeseen fluctuations in demand for the product of the proposed postmining land use may alter the potential profitability of the investment. The net result of these factors is that mining firms have a number of reasons for avoiding the expense of improved postmining land use planning, design, and documentation with the initial permit application, if that expense is of significant magnitude.

The structure of the coal industry in the Appalachian region also tends to inhibit the development of improved uses on reclaimed lands. Primarily, in the majority of cases the mining firm is not the owner of the land being mined. Thus, the firm will not benefit financially from reclamation practices that go beyond the minimum standards. As far as we know, there are no established precedents for contractual arrangements between operators and land owners to require that specified reclamation practices be carried out in exchange for (presumably) a share of the expected profits or benefits of the improved reclamation practice. Even if such contractual arrangements were to be established, they would place additional stress upon the expected profits since both the land owner and the mining firm would require compensation. An additional barrier to the establishment of such arrangements is the current existence of long-term contracts between land owners and mining firms.

The nature of the mining firms and the landowners adds rigidity to the status quo. Mining firms are in the mining business; for the most part, they have little experience with land development for purposes other than mineral extraction. This adds a certain amount of risk to any postmining land development schemes, from the firm's point of view, unless they are willing to bring in outsiders. Given the current low price of coal and the unfavorable financial condition of many firms, additional risk is not of interest to many firms.

The owners of mined lands in the Appalachian region are largely corporate. Since many of the primary land-owning corporations are mainly in the mineral business, they are often unwilling to develop any surface use that will conflict with future deep mining. Given the technologies of today, the presence of housing, for example, on the surface would preclude deep mining of underlying coal due to the potential for subsidence, structural damage, and liability suits.

Another factor to consider is the perception of the mining community of being overregulated, which makes it hesitant to attract regulatory scrutiny through variance application. This attitude is a product of a history of adversarial relationships between the industry and the regulatory agencies. It is also a result of the plethora of regulations governing the coal industry. Environmental performance standards of SMCRA govern many details of operation and are enforced by frequent visits of inspectors to the mining site. These inspectors have the right to impose fines and/or temporarily close the mining site if significant violations are detected.

Finally, there is the effect of remaining technological unknowns. What is the site index of mine soils constructed using controlled overburden placement techniques for southern pine timber production? How does one establish a functional and legally acceptable septic drainfield on a mine soil site? What will be the effect of deep mining on surface subsidence and groundwater quality in the Appalachian coalfields?

VI. CONCLUSIONS

Productive topsoil substitutes can be generated from hard rock overburden in the southern Appalachians, but care must be taken in the selection of the proper strata. It is particularly important to reclamation success that controlled overburden placement techniques be used to generate at least 1 m of loose spoil at the final surface for seeding. Older revegetation practices that were designed to meet a 2-year vegetation bond release period are likely to be ineffective at maintaining a vigorous vegetative community for the 5-year period mandated by SMCRA. The accumulation of organic matter and N over time, and the minimization of P fixation by soil Fe-oxides are critical to long-term site rehabilitation and the future invasion of the site by native woody species.

Another factor that is vital to reclamation success is the necessity to produce stable landforms. Many steeply sloping lands in this region that are returned to AOC are severely limited in their postmining land use by steep final slopes and potential instability. Thus, mining firms have little incentive to invest thought, time, or money in any form of reclamation beyond the minimum standard required by law. These steeply sloping landforms continue to be produced in abundance, in spite of the scarcity of flat lands in central Appalachian coal mining region.

Cooperative university-industry-agency research through the Powell River Project has had a profound effect upon people's attitudes in the Virginia coal mining region. In the past 10 years, tremendous progress has been made towards the goal of widespread, successful reclamation. Knowledge and techniques have been developed which allow mining overburden (spoil) to be transformed into mine soils with productive potential far exceeding the native soils of the region. Knowledge continues to be developed regarding management techniques that will allow carefully constructed mine soils to reach their full productive potential, once the day comes that they are produced in abundance. However, significant obstacles remain. These include legal and institutional constraints to improved reclamation as well as technological unknowns. As a result, the Powell River Project has expanded its research efforts into these areas, with the objective of stimulating widespread application of improved reclamation techniques.

REFERENCES

1. **Daniels, W. L. and Amos, D. F.**, Generating productive topsoil substitutes from hard rock overburden in the Southern Appalachians, *Environ. Geochem. Health,* 7, 8, 1985.
2. **Barnhisel, R. I., Powell, J. L., Akin, S. W., and Ebelhar, M. W.**, Characteristics and reclamation of "acid sulfate" mine spoils, in *Acid Sulfate Weathering,* Soil Science Society of America Spec. Publ. No. 10, American Society of Agronomy, Madison, Wis., 1982.
3. **Rives, C. S., Bajwa, A. E., Liberta, A. E., and Miller, R. W.**, Effects of topsoil storage during surface mining on the viability of VA mycorrhizae, *Soil Sci.,* 129, 253, 1980.
4. **Smith, R. M., Sobek, A. A., Arkle, T., Jr., Sencindiver, J. C., and Freeman, J. R.**, Extensive Overburden Potentials for Soil and Water Quality, U.S. Environmental Protection Agency, 60012-76-184, 1976.
5. **Armiger, W. H., Jones, J. N., Jr., and Bennett, O. L.**, Revegetation of Land Disturbed by Strip Mining of Coal in Appalachia, U.S. Department of Agriculture, ARS-NE-71, 1976.
6. **Berg, W. A.**, Limitations in the use of soil tests on drastically disturbed lands, in *Reclamation of Drastically Disturbed Lands,* American Society of Agronomy, Madison, Wis., 1978, 653.
7. **Howard, J. L.**, Physical, Chemical and Mineralogical Properties of Mine Spoil Derived from the Wise Formation, Buchanan County, Virginia, M.S. thesis, Virginia Polytechnic Institute and State University, Blacksburg, 1979.
8. **Sweeney, L. R.**, Soil Genesis on Relatively Young Surface Mined Lands in Southern West Virginia, M.S. thesis, Virginia Polytechnic Institute and State University, Blacksburg, 1980.
9. **Everett, C. J.**, Effects of Biological Weathering on Mine Soil Genesis and Fertility, Ph.D. dissertation, Virginia Polytechnic Institute and State University, Blacksburg, 1981.
10. **Plass, W. L. and Vogel, W. S.**, Chemical Properties and Particle Size Distribution of 39 Surface Mine Spoils in Southern West Virginia, U.S. Department of Agriculture Forest Service, Res. paper NE-276, N.E. For. Exp. Stn., Upper Darby, Pa., 1973.
11. **Gensheimer, S. J. and Stout, W. L.**, Potassium and magnesium availability in mudstone coal overburdens in West Virginia, *J. Environ. Qual.,* 11(2), 227, 1982.
12. **Daniels, W. L. and Amos, D. F.**, Mapping, characterization and genesis of mine soils on a reclamation research area in Wise County, Virginia, in *Proc. Symp. Surface Mining Hydrology, Sedimentology, and Reclamation,* University of Kentucky, Lexington, 1981, 261.
13. **Marrs, R. H., Roberts, R. D., and Bradshaw, A. D.**, Ecosystem development on reclaimed China Clay wastes. I. Assessment of vegetation and capture of nutrients, *J. Appl. Ecol.,* 17, 709, 1980.
14. **McCart, G. D. and Daniels, W. L.**, Liming and Fertilizing Minesoils, Virginia Cooperative Extension Service Publ. #460-102, Virginia Polytechnic Institute and State University, Blacksburg, 1984.
15. **Slick, B. M. and Curtis, W. R.**, A Guide for the Use of Organic Materials as Mulches in Reclamation of Coal Minesoils in the Eastern United States, U.S. Department of Agriculture Forest Service, N.E. For. Exp. Stn., General Technical Report NE-98, Upper Darby, Pa., 1985.
16. **Brown, M. R., Wolf, D. D., Morse, R. D., and Neal, J. L.**, Viability of rhizobium in fertilizer slurries used for hydroseeding, *J. Environ. Qual.,* 12(3), 388, 1983.

17. **Schoenholtz, S. H. and Burger, J. A.,** Influence of cultural treatments on survival and growth of pines on strip mine sites, *Reclam. Revegetat. Res.,* 4, 223, 1984.
18. **Dancer, W. S., Handley, J. F., and Bradshaw, A. D.,** Nitrogen accumulation in Kaolin mining wastes in Cornwall. I. Natural communities, *Plant Soil,* 48, 153, 1977.
19. **Dancer, W. S., Handley, J. F., and Bradshaw, A. D.,** Nitrogen accumulation in Kaolin mining wastes in Cornwall. II. Forage legumes, *Plant Soil,* 48, 303, 1977.
20. **Ashby, W. C., Vogel, W. G., and Rogers, N. F.,** Black Locust in the Reclamation Equation, U.S. Department of Agriculture Forest Service, N.E. For. Exp. Stn., Report NE-105, Upper Darby, Pa., 1985.
21. **Daniels, W. L., Bell, J. C., Amos, D.F., and McCart, G. D.,** First year effects of rock type and surface treatments on mine soil properties and plant growth, in *Proc. Symp. Surface Mining Hydrology, Sedimentology, and Reclamation,* University of Kentucky, Lexington, 1983, 275.
22. **Sopper, W. E. and Seaker, E. M.,** Strip Mine Reclamation with Municipal Sludge, U.S. Environmental Protection Agency, EPA-600/S2-84-035, Municipal Environmental Research Laboratory, Cincinnati, 1984.
23. **Bergerson, F. J.,** Biochemistry of symbiotic nitrogen fixation in legumes, *Am. Rev. Plant. Phys.,* 22, 121, 1971.
24. **Bell, J. C. and Daniels, W. L.,** Four case studies of slope stability on surface mined lands returned to approximate original contour in SW Virginia, in *Proc. Symp. Surface Mining Hydrology, Sedimentology, and Reclamation,* University of Kentucky, Lexington, 1985, 237.
25. **Zipper, C. E. and Daniels, W. L.,** Economic monitoring of a contour surface mine in steep slope Appalachian topography, in Proc. *Symp. Surface Mining Hydrology, Sedimentology, and Reclamation,* University of Kentucky, Lexington, 1984, 97.
26. **Zipper, C. E., Hall, A. T., and Daniels, W. L.,** Costs of mining and reclamation at a contour surface mine in steep slope topography, in *Proc. Symp. Surface Mining Hydrology, Sedimentology, and Reclamation,* University of Kentucky, Lexington, 1985, 193.
27. **Zipper, C. E., Chakraborty, A., Topuz, E., and Daniels, W. L.,** A surface mining simulator for application in steep slope topography, in *Proc. Symp. Surface Mining Hydrology, Sedimentology, and Reclamation,* University of Kentucky, Lexington, 1985, 25.
28. **Zipper, C. E. and Daniels, W. L.,** COSTUM: A System for Analysis of Operational Cost Data from Coal Surface Mines, Virginia Agricultural Experiment Station Bulletin 86-1, Blacksburg, 1986.
29. **Radian Corp.,** CPS/PC: Advanced Software System for Gridding, Contouring, Mapping, and Analysis, 8501 Mopac Blvd., Austin, Tex., 1985.
30. **Zipper, C. E.,** Opportunities for Improved Surface Coal Mine Reclamation in the Central Appalachian Coal Fields, Ph.D. dissertation, Virginia Polytechnic Institute and State University, Blacksburg, 1986.
31. **Zipper, C. E., Bell, J. C., and Daniels, W. L.,** A detailed study of the stability, land use potential, and erosivity of surface mined lands returned to their approximate original contours, in *Proc. Powell River Project Symp.,* Virginia Agricultural Experiment Station, Blacksburg, 1986, 38.

Chapter 10

RESTORATION AND MANAGEMENT OF ECOSYSTEMS FOR NATURE CONSERVATION IN WEST GERMANY

Diedrich Bruns

TABLE OF CONTENTS

I.   Introduction...164

II.   The Problem: Loss of Certain Species and Ecosystems........................164

III.   Social Background and Legal Framework......................................165

IV.   Ecosystems Which Cannot be Restored..166

V.   Methods..167
        A.   Restoration by Transplantation.....................................167
                1.   Problem..167
                2.   Example 1..168
                3.   Example 2..168
        B.   Restoration by Implantation (Partial Transplantation)..............169
                1.   Problem..169
                2.   Example 1..169
                3.   Example 2..169
        C.   Restoration by Natural Colonization................................170
                1.   Problem..170
                2.   Example..170
        D.   Restoration by Regulation of the Water Regime......................171
                1.   Problem..171
                2.   Example 1..172
                3.   Example 2..172
        E.   Restoration by Reclamation of Flooded Gravel Pits for Lake
                Development..172
                1.   Problem..172
                2.   Examples...172
        F.   Restoration by Removal of Biomass..................................173
                1.   Problem..173
                2.   Example 1..173
                3.   Example 2..174
                4.   Example 3..174
                5.   Example 4..175

VI.   Other Restoration Methods..175

VII.   Case Studies...176
        A.   Case Study I — Restoration of the Fliede-Fulda Marsh...............176
        B.   Case Study II — Restoration of a Creek Ecosystem..................177
        C.   General Observations...180

VIII. Discussion 182

IX. Conclusions 183

Acknowledgments 183

References 184

I. INTRODUCTION

Restoration of disturbed land and water has a long tradition in West Germany. Even with the construction of the first Autobahnen, ecological factors in the existing landscape were taken into consideration and special seed mixtures were chosen for seeding the highway embankments. The reclamation efforts of the mining industry on the Ruhr and in Cologne Lignite District are world renowned for their advanced and highly specialized methods.[1,2] The technique of using plants as living "building materials" is widely respected.[3] Until the present, restoration efforts were directed primarily towards the installment of usable agricultural land, the creation of timber stands, and large scale recreation parks. Another traditional restoration goal is the clean up of polluted lakes and rivers.[4]

Most recently, the restoration of ecosystems as habitat for indigenous animals and plants is rapidly gaining importance under the specific ecological and socio-legal conditions and central Europe. As a result of massive losses of certain species and ecosystems, public pressure is turning towards both conservation and restoration, and new laws have been passed. For a better understanding of the presentation and discussion of concepts and methods, an explanation of the required conditions and reasons for the restoration of ecosystems in Germany is included in this discussion.

A discussion of all management options for the restoration of ecosystems, all standard methods and certainly all cases of impacts is beyond the scope of this chapter. I have limited information to my experience in case studies pertinent to conservation and restoration of ecosystems as habitat at the present time.

II. THE PROBLEM: LOSS OF CERTAIN SPECIES AND ECOSYSTEMS

Centuries of man's interactions with nature have created an extraordinary diversity of ecosystems in central Europe. The casual visitor to West Germany still receives a superficial impression of rich, diverse, and intimate landscapes. Unfortunately the cultural landscape of bygone days no longer exists. "Over-cultivation"[5] is rapidly reducing the heritage landscape to a "biological desert".[6]

The long period of human habitation in Europe has resulted in the alteration of virtually every existing ecosystem.[7] Species once associated were regrouped, continuously adding new species. New habitat types were created, such as pastures and fields. This process has continued until the present time with mines, quarries, railroads, and other industrial habitats presenting more recent opportunities for formation of new species associations.[8]

However, the speed with which existing ecosystems disappear and new ones appear has changed. Environmental conditions are eliminated before species have time to disperse to distant habitats. The speed with which man changes his present environment precludes adaptation.[9,10]

Table 1
TOTAL NUMBER OF VASCULAR PLANTS AND
VERTEBRATE SPECIES (SELECTED TAXA)
AND NUMBER OF EXTINCT AND
THREATENED SPECIES IN WEST GERMANY[12]

		Extinct and threatened	
Taxon	Total number	Number	Percent of total
Vascular plants	2476	697	28
Mammals	94	44	47
Birds	305	98	38
Reptiles	12	9	75
Amphibians	19	11	58
Freshwater fish	70	49	70

One may be tempted to invoke the European high population density as an explanation for the intensive land use, but the reasons are mainly political. The guaranteed farming subsidies for cash crops make it attractive to produce large crop surpluses independent of market prices. This encourages high agricultural land turnover and heavy fertilization. In addition to this, large scale land consolidation has changed many regions of the "picturesque landscape" into large tracts of monotonous farmland. The old villages and towns look out of place.

These changes have resulted in the extinction of many rare species. Many common species have become rare. Taking invertebrates into account, it has been estimated that West Germany is presently losing approximately 90 species annually.[11] More than 50% of the total number of indigenous species is either extinct or threatened (Table 1).[12]

To quantify which ecosystems are associated with extinct and threatened vascular plants, an extensive analysis was conducted.[13] The results are in accordance with the national inventories of habitats worthy of protection.[14] Certain rare habitat types are associated with significant numbers of extinct and threatened species. These habitats also demonstrate a consistent pattern of ecological extremes with severe conditions of exposed and arid sites, extremes of moisture, and harsh infertile soil of poor nutrient availability. These qualities are disappearing rapidly from the central European landscape. Prominent examples are peat bogs ("raised bogs") and other wetlands of low and medium fertility (oligotrophic and mesotrophic wet meadows, fens, lakes and ponds); natural streamside (riparian) habitats; freshwater mud flats; vegetation dependent on traditional farming practices such as dry grasslands, heaths and field communities; and arid open woodland and rock habitats.

III. SOCIAL BACKGROUND AND LEGAL FRAMEWORK

A new social consciousness has arisen in West Germany and in neighboring countries in the past 2 decades. The complexities of this phenomenon are fascinating and are probably best mirrored in the meteoric rise of the environmentalist "Green Party" in its political significance. The impossibility of continuing the postwar urban sprawl in West Germany has become apparent, and the problems associated with unlimited technological development have caused many people to reassess their natural heritage as a positive thing to enjoy and preserve. The leisure time necessary for appreciation of the outdoors has been provided by the economic wonder of the postwar boom, financial security, and economic and social stability to a degree unparalleled anywhere else in the world today. Unfortunately, there is little in the way of natural lands remaining to enjoy. Hence, natural, and even historic or heritage landscapes, have, in the last few years, increased considerably in political value.

The great loss of species and ecosystems through relentless human intervention has also penetrated the public consciousness. Increasing numbers of people are dedicating time and money to the preservation of a particular species. This is popularized through actions like "Bird of the Year". Groups of species receive special attention in the form of assistance programs, primarily initiated by private interest groups, e.g., The German Society for the Protection of Birds (Deutscher Bund Für Vogelschutz). Many an amateur naturalist has acquired extensive knowledge of a small specialty. Understanding of ecological relationships has not developed in parallel so that a unified approach to nature conservation is missing. This is especially pertinent with reference to "static" and "dynamic" *philosophies of nature conservation.*

The static approach strives toward preservation of the status quo. By primarily focusing on endangered species, certain areas are painstakingly preserved as habitat. Often they are managed with investment of considerable time and energy, e.g., through removal of shrub growth on pastures, with respect to a few species of special interest. As long as many of the dynamic qualities of natural and man-made ecosystems as well as the capacity of certain species to adapt to changing environmental conditions are not understood, the static approach of nature conservation is valid.

The dynamic approach sees changes in the landscape as one important prerequisite for the preservation of species diversity even within the cultural landscape. If we do not wish to change our environment into a museum, we must logically permit a changing and dynamic landscape. We must, however, exercise control over these changes. In particular, we need to learn to direct and regulate impacts. This also means that we must learn to admit that certain impacts are so damaging that they cannot be permitted to occur. Impacts that can be incorporated into sound environmental management are those that allow a preexisting system to buffer the changes without resulting in a total destruction of the previous biological and physical equilibrium.

Despite heterogeneous ecological philosophies, it has been possible, because of present social pressures, to pass a federal conservation law.[15] The most important statement in the law is that serious impacts on the special character of the countryside and disturbances in natural cycles and processes must be avoided. The glaring weakness and obvious paradox in the law follows immediately and states that, should the avoidance of impact not be possible (due to political decisions in favor of other social values, i.e., roads), unavoidable disruption must be mitigated or compensated for. A major failing was the passing of this law prior to the ecological interpretation of the legal term "compensation". Even now, after 9 years, little progress has been made in the clarification of this suspiciously vague term. In 1985, the European Economic Community (EEC) passed directives on environmental impact assessment requiring all member-states to pass national legislation by 1988. This will force the West Germany legislators to clarify and revise the existing legislation to meet standards proposed by the EEC.[16]

IV. ECOSYSTEMS WHICH CANNOT BE RESTORED

The archaic conservation of the landscape is unrealistic. This contradicts the natural flux. Many species are dependent on constant renewal. The more dynamic the flux in an ecosystem is, the shorter is the cycle or renewal and the greater the possibility of success of reclamation efforts.

On the other hand, there are ecosystems which are, under present conditions, not restorable.[17] These include, as shown in Table 2,

1.   Primary ecosystems
2.   All ecosystems that developed under human influence where the conditions causing

Table 2
ECOSYSTEMS THAT CANNOT BE RESTORED IN WEST GERMANY
(EXCLUDING THE ALPS, NORTH SEA TIDAL FLATS, AND URBAN
SETTLEMENTS) ACCORDING TO KAULE IN 1986[18]

Primary ecosystems	Raised bogs, including bog fragments
	Mesotrophic and oligotrophic fens and carrs
	Primary lake ecosystems, including natural lake shores
	Natural and seminatural sections of streams, including flood plains
	Primary rock ecosystems
	Salt marshes and dunes
Ecosystems which developed under conditions no longer in existence, or those needing long recovery periods	Floodplains with meadows, riparian forests, and marshes
	Old secondary rock ecosystems
	Old heath ecosystems
	Salt meadows
	Protected forests
	Outstanding features in the landscape

Ecosystems in which species listed in the "Red Data Books" are present in isolated colonies; successful recolonization of equivalent habitats is unlikely.
Ecosystems of national and international importance.

this development cannot be recreated, or where the ecosystem requires too long an evolutionary period

3. All ecosystems, including those developed under human influence, in which species listed in the Red Data Book are present in isolated colonies, in which a recovery period is too long, or successful recolonization is too unlikely

From these generalizations, a list of "Taboo ecosystems" evolves. These must be preserved and will not be considered further. Other environmental conditions would result in different lists (Table 2).[18] The Alps and coastal mud flats are special cases. In addition, connected habitat complexes in central Europe that contain endangered species requiring large territories should be protected. These include complete and undissected drainage areas and large forests.

The Taboo-habitats described in Table 2 are definitive. It is possible that other types can be added to these lists. Habitats not described have still to be analyzed with reference to their suitability for restoration under present central European conditions. In this context, one must take into consideration that, despite preservation of the habitat status quo in the next few years, many species will regionally and nationally vanish and be unavailable for recolonization. This emphasizes the urgency of a habitat renewal program. The longer this is postponed, the less change there will be for success. This is especially significant when considering that prolonged maintenance of the *status quo* will not be possible.

V. METHODS

A. Restoration by Transplantation
1. Problem

Because of high population density and intensive land use, the destruction of certain habitats cannot, despite extensive planning, always be avoided. Large projects such as airports, roads, railways, and land consolidation projects are particularly implicated. Various experiments of the last few years have shown that it is possible to transplant certain ecosystems as an entity, or in parts. These methods have been used for a long time during the construction of botanical gardens. This is only successful with certain habitat types. The cost is extremely high and can only be seldom financed. For the Floralies Internationales 1976 and at the Internationale Gartenbau Ausstellung 1983, bogs were dug up, transported, and replanted.[19,20] In both cases, the habitats changed considerably and the limits of this method become quickly apparent. Transplantation of hedges has been successful. A new hedge requires a couple hundred years for the soil, water regime, and organisms to reestablish an equivalent diversified ecosystem.[21] Linear elements such as hedges are amenable to transplantation, and this is the method of choice for such a situation.

2. Example 1

In West Germany, the extensive land reform and consolidation procedures have a special place in the destruction of habitats in historic landscapes. The purpose of land consolidation is to produce the largest possible agricultural areas by homogenization of the landscape. Hedges are in the way. Legally, the land consolidation act of 1976 requires the preservation of wooded areas so that projects such as the transplanation of hedges were tried very early.[22,23] A large ditch is dug in an area with similar geological, hydrological, and microclimatic conditions. The hedge is cut back. Sections of hedge are cut and transferred immediately by loaders to the ditch. No intermediate vehicles, such as loading onto trucks, are used. Sections would disintegrate should there be too many intermediate transfer steps. Spaces between sections are filled with soil lacking humus (skeletal soils). Depending on topography and season, irrigation is advantageous. During transplantation, a representative number of the typical plants and animals are rescued so that the house cleaning effect of land consolidation is minimized. A problem not yet solved is what to do with hedges containing old thick trunks. Equally unsuccessful is the transplantation of stone wall hedges typical for many regions. These structures do not contain transplantable units. Here, new developments through primary succession on loose stone and skeletal soils may be more successful: an attempt at mimicking the ancient process.

A problem of transplantation is the release of nutrients. Oxygen penetrates cracks between transplanted sections, mineralizing organic materials and making them available as nutrients. In this fashion, certain competitive species are given an advantage. This problem can be minimized by cutting large sections and by carefully filling open areas. For ecosystems very sensitive to nutrient release (raised bogs), transplantation has not been very successful, as example 2 shows.

3. Example 2

During the airport expansion project, Zürich-Kloten, parts of oligotrophic wet meadows, mowing marsh (sedgemarshes created by hundreds of years of regular mowing, associated with rare orchids) were transplanted in order to preserve them. This bog was the only one of its kind in Switzerland. The target site and the donor site were equivalent with respect to water regime and nutrient levels. Groundwater level could be regulated. The target site was surrounded by equivalent vegetation: wet meadows and swamp forest relicts. However, in the new habitat, plants are appearing that suggest eutrophic conditions (e.g., *Solidago gigantea*).

Special vehicles were developed for this project to cut and transport sods of exactly 50 cm depth, and 90 × 130 cm in surface area. In the target area, sods of the same size were removed. Only short transport distances were necessary. The method does not work for longer distances. The planted areas were protected with vertical flat tiles to prevent root invasion from the surrounding area.[10,24,25] Despite careful technique, considerable and significant ecological changes occurred. During transport, soils were lost, exposing roots and creating an uneven surface resulting in differing moisture conditions. During cutting of sods, deep roots were severed, so that certain plants were systematically damaged or eliminated.

Joints and tears in the sods could not be completely sealed, so that fen peat mineralized on the entry of oxygen. Nutrients were released and invading plants quickly established themselves. These larger plants shaded light-dependent species and changed the milieu. The biomass production increased substantially. Despite directed efforts at annual harvest of biomass, susceptible species were eliminated.[26] Even though the nutrient content of the groundwater is unchanged, the vegetation is now reminiscent of a eutrophic wet meadow with predominance of *Filipedula ulmaria, Lysimachia vulgaris, Calamagrostis epigeios, Cirsium palustre, Angelica sylvestris, Juncus subnodulosus,* and other species common in the surrounding area. Saving the original bog was not possible. The value of the experiment according to Klötzli[26] (translation by author) was that " . . . it is important to learn how much effort is involved in removing a fresco of nature, in restoring the fresco in a suitable new home and how impossible it is to reconstruct a Mona Lisa of nature in its entirety."

B. Restoration by Implantation (Partial Transplantation)

1. Problem

As described previously the destruction of habitats cannot always be prevented. The transplantation of ecosystems has narrow limits and requires a search for other methods to save endangered species. To bypass the problem of eutrophication in particular, methods were developed of transplanting parts of ecosystems, which contain the crucial species but only the least amount of organic material. This involves artificially directed primary succession through a process of innoculation of skeletal soils. For example, only a few sods containing seeds, plant parts, and soil organisms are used to cover some small portions of the restoration site. This method is successful when a specific trophic level[27] is to be preserved. The more oligotrophic the recipient site is to be, the more difficult the application of this method. If, in the neighboring areas of the recipient site, there are ecosystems compatible with the goal community, this method of restoration should be used.

2. Example 1

One of the first applications of the implantation method began in 1983 during the construction of the Rapid Railway Project described in further detail as a case study later. For the restoration of the wet lands to be destroyed, it was necessary to establish several trophic levels in the new ponds. Nutrient poor conditions were achieved through pond excavation in which all top solid and organic material were removed. Nutrient rich sites were produced through implantation of fertile material into prepared recipient sites from donor ponds.

Long before the existing wetland was destroyed, the recipient sites were constructed. A dredger cut pond embankment sections from the condemned donor wetlands and placed them in the new ponds in the same order but slightly lower in relation to the water level. Typical marsh and reed species dominate. Because many animal and insect species hibernate in the pond mud, this material was also implanted on the new sites before the final water level was reached.

There has been ongoing observation of the project since initiation of the restoration measures. All plant species found and inventoried prior to habitat destruction have reappeared in the recipient site. In addition, some rare pioneer plants have also appeared.

3. Example 2

The largest restoration project for nature conservation purposes in West Germany (2.5 million DM or ~1.0 million U.S. $) using the implantation method was undertaken because of the construction of the Rhein-Main-Danube Canal. The river Danube is presently undergoing remodeling into a major cargo route through central Europe. Near Regensburg, 54 ha of riparian forest, sedge marshes, lakes, ponds, and 48 ha of periodically flooded grassland had to be destroyed. It was not possible to transplant the entire complex. Large sods containing rare plants were removed and transplanted. In 1983, before the transfer, the vegetation and fen peat were analyzed to permit as close a reduplication as possible on the recipient new site. Sods were removed and stored until preparation of the new flood embankments was complete. For differing plant communities, different intermediate storage solutions were devised. Irrigation back up systems kept fastidious plants moist. The stored plants rested on a base of prepared lake mud. Sods of marsh reeds and submerged vegetation were excavated with dredgers and "planted" with loaders. Both machines were equipped with especially large shovels to prevent sod breakage and to minimize the number and size of joints. Some sods had a base of 5 m². The intermediate storage area was complete in February 1984 and was surrounded by a protective fence. The 20,000 m² of sods and 10,000 m³ of peat and mud were stored at the intermediate site. In January 1985, the recipient site had been completed and the sods were transferred from the storage facility. Some vegetation types were implanted on skeletal soils and others onto peat and mud. An attempt to imitate the original stratification of the river bank was made. The sods were therefore placed in staggered rows and the spaces between were filled with peat and mud. Natural colonization is to occur in the spaces between. Shrubs and bushes were also planted in some areas.[28]

Conclusions are not yet possible. It remains to be seen whether the desired species can disperse rapidly enough to prevent the inevitable weed growth that will occur on the nutrient rich peat and mud used to fill in the "spaces" between implanted sods. This problem will be of varying severity in the different strata of the new embankment.

C. Restoration by Natural Colonization
1. Problem

If the methods of transplantation and implantation seem to be inapplicable, primary succession is another option. The goal community will not be immediately available in its final form. However, spontaneous and uninhibited succession in the cultural landscape is extremely rare. Many pioneer species with high light needs and low nutrient tolerance are endangered. The resultant creation of earlier succession stages of an ecosystem-type not only helps achieve the goal-community, but also provides a habitat for many generally endangered species. The development of an ecosystem on skeletal soils (including porous rock) is a natural part of the evolutionary process. Examples are dunes, landslides, river bank erosion, new gravel banks, glaciation, and volcanic activity, only to name a few. Pioneer species settle the areas. They are displaced by species of subsequent succession steps, more or less in an orderly fashion.[29-32] The path of succession is determined by the choice of adjacent ecosystems. This could be clearly demonstrated through analysis of clay pits in the Stuttgart area.[33] Areas of skeletal soils adjoining natural ecosystems are rapidly invaded and populated in accordance with the species diversity of neighboring ecosystems.[33,34] With directed preparation of the new areas for certain ecosystems, only a few years will pass before a complete cross-section of species from neighboring habitats is present. It is important to prevent the appearance of ubiquitous species, that under conditions of nutrient availability (especially nitrogen) are successfully competitive and prevent immigration of the desired species.

2. Example

There are few examples of the use of succession for the deliberate development of ecosystems on skeletal soils. The method was used successfully by the author and appears in Case Study I. The unintentional evolution of stable habitats on skeletal soils can be studied very well in old quarries and similar places. Bradshaw and Chadwick[1] remarked "that of the 3000 sites recognized officially as sites of special scientific interest in England and Wales, 75 are quarries and other mineral workings." They continue: "Since old quarries and similar spaces are empty, the soil and plants have been removed and all that remains is bare ground and rock, this provides a refuge for plants that cannot stand the competition of the more vigorous plants that grow on good soil.'' Not only do unreclaimed areas become refuges for certain species but '' . . . because they are left alone, they can also become wilderness places where the various stages of development of ordinary plant communities can be seen, the open grassy vegetation which will lead to mature grassland, the scrub that will lead to woodland.'' The development of ecosystems on skeletal soils in old quarries has been extensively investigated. Their potential significance of protection and reconstruction of species diversity is well documented.[35,36] Gravel pits are also in this category.[37,38] Less well investigated are sand pits, clay pits,[33,39] old rock, and gravel embankments.[40,41]

Not all reclaimed extraction sites develop into habitats worthy of protection. By closer examination of documented cases, the importance of the source pool of species in adjoining habitats becomes apparent. Similar results were obtained during a study of naturally colonized waysides on the stony soil of the Swabian Jura.[34] ''Only if the nearest source of appropriate immigrants is at a distance and nutrient levels on the site remain low, then colonisation will be limited to those species with means of long distance dispersal.''[42]

D. Restoration by Regulation of the Water Regime
1. Problem

A difficult situation is presented by the falling groundwater table secondary to drinking water wells, construction, and peat mining. Temporary or sometimes permanent compensatory solutions can be occasionally obtained through irrigation and damming of the surface water. Ecosystems dependent on certain water levels can be artificially supported in this fashion. Irrigation via a pump system is, in this setting, complicated, expensive, and an unsatisfactory permanent solution. In combination with the transplantation method, irrigation can be used to preserve sods during the construction period as described in example 2 under restoration by implantation. For the preservation of wetlands overlying drinking water wells, the damming of a drainage area can be of use. This method has proved useful in flat areas where minimal efforts in damming are required to provide the sensitive ecosystem with the necessary water supply. Problems appear when periodic changes in water quality and quantity are sufficient to endanger the restoration project. In central Europe the damming of a drainage system can lead to the destruction of the last intact stream or river of a region. In this situation, the cost-benefit relationship of the measures must be taken into account. In the past few years, many ponds for endangered amphibians have been created with the help of drainage system damming. The destruction of intact creeks for the creation of frog ponds should not be the goal of ecosystem restoration.

The damming of surface water in abandoned peat mines permits the evolution of secondary bog ecosystems. This does not recreate ancient raised bogs. However, the vegetation and many animals of fen ecosystems can settle again in flooded open cast peat cuttings and nutrient poor gravel pits. Although the flooding of bogs has been pursued on a small scale in West Germany to this end, the most well-known example for the success of this is the Norfolk Broads in the middle of East Anglia, U.K., the remains of a medieval peat industry.[1] Active attempts at the restoration of bog and muskeg ecosystems can orient their approach to secondary bogs and the observed processes within them. Open water communities, reed swamps, marshes, and special fen ecosystems have, through natural succession or through artificial damming, developed. In eutrophic and mesotrophic milieus of certain wet old fields, fen-like communities, sedge, and rush marshes and reeds have grown. Oligotrophic blanket fens develop near wet gravel beds in subalpine, gravel rich rivers, and in gravel pits when species are found in directly adjacent communities. A well-documented example is a secondary blanket fen in a gravel pit south of Ingolstadt.[43] Experimental restoration efforts of bog and muskeg ecosystems are found in Eigner and Schmatzler[44] and Schwaar.[45] Biological sewage disposal systems can develop eutrophic marshes and fens.[46]

2. Example 1

South of the urban conglomeration and metropolitan complex of Mannheim-Ludwig-shafen is an as of yet undisturbed natural groundwater storage basin. According to prediction, the use of this aquifer will result in a change in the surface water in certain areas. A drop in the water table is especially dangerous for wetland ecosystems in the old river channels of the Rhein. An artificial water supply was suggested as a compensatory measure to prevent the drop of the water table. For this, special trickle-irrigation systems were designed, where the location of the trickle and the amount of necessary water were calculated through systematic alternative analysis. The trickle flows from the lower terrace of the Rhein Valley where the pumped water is channeled over already existing farm ditches. The complete installation will be a combination of rivers, dams, ditches, and pump systems.

3. Example 2

Damming of rivers for the regulation of river erosion or for drinking water causes considerable changes in the water regime. The water level is raised in the immediate area of the dam and the water stops moving. Hybrid waters develop: no longer rivers but not yet lakes. The periodic flooding of the natural riparian forests no longer occurs. River sediments accumulate behind the dams. Examples of this are described for the lower River Inn on the German-Austrian border,[47] for the river Reuss is the Swiss Midlands,[48] and for the Isar River north of Munich.[49] In none of the above cases was it possible to preserve the typical riparian ecosystem. To make the best of the situation, eco-technical measures were instituted to change the neighboring ecosystems into wetland habitats. They can be used as examples for this. Old arms or ox bows have been newly flooded and are becoming eutrophic stagnant water. Gravel bars and gravel and sand banks of varying slopes are available as skeletal soils for primary succession. Permanently flooded riparian forest is changing into swamp forest. Silt and gravel accumulation assures new formation of shallow water habitats. This also occurs through flooding of former riparian bottomland.

E. Restoration by Reclamation of Flooded Gravel Pits for Lake Development

1. Problem

Resulting from widespread pollution, the natural aging process of many lakes and ponds is accelerated through eutrophication by a factor of 10 to 20.[18] Nutrient rich lakes should therefore not be included as a goal for the restoration of ecosystems, unless locally endangered species are at stake. With the help of specialized technology, many eutrophicated waters can be cleaned.[4] In addition to cleaning contaminated lakes, the creation of oligotrophic lakes in gravel and sand excavations for purposes of nature conservation is taking on increasing importance.[50] If nutrients can be successfully kept at bay, it is possible, with a minimum of energy, to establish necessary living conditions for native plants and wild animals that have become rare secondary to the disappearance of this type of habitat. These species clean stagnant surface water.[43,51]

2. Examples

One of the most well-known examples is the Sevenoaks gravel pit.[52] A summary of the most important ideas and examples for converting gravel pits into wildlife preserves in Germany can be found in Dingethal et al.[53] Dahl and Jürging[37] describe many successful projects of reclamation of gravel pits for conservation purposes. Prerequisite for the successful reclamation was that the excavation of gravel proceed in a fashion compatible with the goal of nature protection so as to achieve a desired morphology and distribution of substrates and skeletal soils. Special habitats for certain species were prepared and included waterfowl nesting sites (such as gravel banks and vertical banks) and waterfowl feeding sites (such as shallow water habitats and reed fields). Although no planting was done, in the first summer, water plant communities and reeds developed due to natural colonization. The first successional phase of pioneer plants was displaced by more stable communities already in the 3rd year. The waterfowl expected to invade the new site arrived almost instantly. Parallels to the Sevenoaks sites are striking, although in the German examples, no plantings or management were undertaken.

F. Restoration by Removal of Biomass

1. Problem

Centuries of site-adapted farming techniques have created a variety of different grassland types. Like all other cultural ecosystems (plagioclimax communities), old, historic, grassland ecosystems require continued management to exist. Because traditional uses are in many cases no longer cost-effective, these grasslands are slowly being lost. They are disturbed by drainage, ploughing, and reseeding or "renovated" with heavy fertilizer or herbicides. Many are lost due to abandonment and secondary succession.

Although historic grasslands are not strictly natural communities " . . . it is probable that they are close analogues of quite natural ecosystems maintained before the advent of agriculture to Britain (and central Europe, author's addition) by wild cattle *Bostaurus*, wild horses *Equus* sp. and other native herbivores."[54]

The most significant factor causing the loss of species on the continental grasslands is the nutrient load. A secondary factor of considerably less importance is physical damage of mowing, grazing, and trampling. On the other hand, up to 40 kg of nitrogen per hectare per year precipitates in some areas of West Germany, and every abandoned field is subject to fertilizer drift from areas of intensive agriculture.[18] On the other hand, biomass is no longer harvested from abandoned fields resulting in accumulation of nutrients on these traditionally oligotrophic areas.

Secondary succession in old grasslands can lead to the development of interesting and valuable habitats in themselves. One must be careful not to destroy one useful habitat for another.

As with all habitat types in West Germany, it is the ecological extremes that are most endangered. These are the nutrient poor wet grasslands and dry grasslands. These can be classified according to plant communities.[55] Famous plant communities in southern Germany are wild flower rich limestone "juniper heath", a parklike grassland on eroded hillsides and hummocks, and the equally spectacular "Streuwiesen", which are wet grasslands on lime containing fen peat and were traditionally sources of stable litter for farm animals. The restoration of these and similar communities is done in several stages according to the progress of agricultural improvement or secondary succession. These steps include the removal of trees and shrubs, intensive mowing, and removal of biomass and continued regular mowing, grazing, or burning. These last three measures continue to remove nutrients after the impact of the initial intervention. Varying techniques are illustrated with the following examples.

2. Example 1

Many typical sheep fescue turfs and the wildflower rich chalk grasslands (dominated by *Brachypodium pinnatum* or *Bromus erectus,* respectively) on the thin calcareous soils of the south German jurassic escarpments (Swabian Jura) are now abandoned due to low profitability of traditional sheep farming methods. The subsequent secondary succession is accompanied by the appearance of calciole herbs. The shrub stage is quickly reached. Rodi[56] describes experiments involving shrub removal to restore the original grasslands. Old (1956) and new aerial photographs are compared and vegetation analysis serves to identify species rich grasslands that are amenable to brush cleaning. For comparison, a detailed inventory of experimental plots was performed. Areas with diverse succession stages served as a departure point.

On first glance, there are many similarities between primary grassland communities on Rendzina or Pararendzina soil (Orthids and Rendolls according to U.S. classification) and the secondary anthropogenic grassland communities. The first results of the exact vegetation inventory were to identify these and separate them from one another. The second result was the determination of the speed with which secondary succession was occurring. It became clear in what time periods brush cleaning would be necessary. Dry, low yield grasslands do not need to be mown yearly. The removal of scrub brush is sufficient as a restoration measure. The mown and unmown fields of this kind are hardly distinguishable. During brush cleaning, the regionally characteristic junipers *Juniperus communis* were left and all other trees and shrubs removed. Some fields in various stages of succession were left as controls. The studies are ongoing.

3. Example 2

Grasslands that have been altered through agricultural improvement can be restored by simple measure of biomass (nutrient) removal. Experimental attempts with different management forms have been conducted in the state of Baden-Wurtemberg since 1950. This entails nutrient extraction through regular mowing without fertilizing. These are attempts to restore over-fertilized meadows to the earlier low yield grassland, rich in species diversity. Schiefer[57] discusses the results. Extensive fertilization changes low yield grassland into high yield grassland in a few short years. A change into low yield grasslands through nutrient removal is only possible in regions of minimal soil moisture and naturally low nutrient availability. Suitable areas can be detected through comprehensive ecological inventory. Time required for nutrient removal (nutrient drainage) varies between 3 to 10 years, depending on the initial conditions. Only after nutrient drainage is complete can the specialized species colonize the new habitat (5 to 20 years). The natural recolonization of appropriate species is very slow and often incomplete. On one hand, the low and slowly growing plants are competitively at a disadvantage. They can only disperse when stronger species have disappeared. On the other hand, the seeds of appropriate species occur only in the soils having supported the desired ecosystem at an earlier date. In cases where species must invade from distant sources, recolonization will take a long time. This process can be accelerated by artificial seeding.[58] Some species have low reproductive capacity and require centuries for reestablishment.[58,60]

4. Example 3

Parallel to the experiments described in example 2, mulching trials were carried out. With this management variant of mowing and leaving the cuttings, nutrient extraction was still possible and plants preferring oligotrophic habitats proliferated. An explanation of this phenomenon does not exist at this time. It was observed that there was a relationship between the increase in species diversity and the mulching time and mulching frequency. Early mulching resulted in the rapid decomposition of the organic material, which promoted the proliferation of low growing and poorly competitive species. Late mulching produced a cover of organic material lasting until the following spring, killing certain species and resulting in a patchy vegetation cover. In tall oat grass meadows *Arrhenatherum elatius* and golden oat meadows *Trisetum flavescens*, an early mulch in June is necessary to preserve rich stands of native wild flowers. Semiarid grasslands, "Borstgrasrasen" (dry grassland on acid soil) and "Pfeifengraswiesen" *(Molinia caerulea*, meadows on nutrient poor fen peat) need mulching only every 2 to 3 years, the semiarid grassland in July, and the "Borstgrasrasen" and "Pfeifengrasrasen" in August. It must be stressed that Schiefer's[57] findings not be taken as patent recipes, for every ecosystem has its peculiarities. The changes resulting from mulching are a product of the existing communities and the land use prior to institution of these measures.

5. *Example 4*

Controlled winter burning is believed to be the cheapest and most useful management tool for dry grassland ecosystems.[54] However, the use of this tool requires skill to control fire safely and used too frequently on the same site will considerably reduce animal populations, especially invertebrates. Common practice in the U.S. for decades, pyro-ecological research in Germany is only in its infancy. While fire undoubtedly played a role in changing the prehistoric forest landscape into an agricultural landscape, this tool is without application in present agricultural central Europe. Only strips of grass between fields abundant in wild flowers are burned as suspected breeding grounds of innumerable ''pests''. This occurs without knowledge of pyro-ecology so that the result is usually more damage than help for nature conservation. For these reasons, fire has acquired a bad name, and it is therefore extremely difficult to use as a management tool for nature conservation. Advances in this field are minimal. Techniques and methods are primarily adopted from the U.S. Their use in West Germany is described by Riess[61] and Schiefer.[62] Of the many varieties of typical prairie biomes of the varying grassland ecosystems of continental Europe, only very specific types lend themselves to restoration or preservation through burning. The elimination of the scrub brush of secondary succession does not work. Success has been achieved in such grassland ecosystems where species rich in lignin, and in cellulose, such as grasses and sedges, predominate and in which harvest is impaired by dryness or excessive moisture, or by decomposition of the straw. These include ''Kalk magerrasen'' (dry calcerous grassland), ''Pfeifengraswiesen'', ''Seggenrieder'', and ''Schilfrönrricht'' (sedge marshes and reeds). While reeds and marshes seldom require such management for preservation, the first two mentioned grassland types are amenable to use of controlled winter burning. Their characteristic species have died back by October *(Bromus erectus, Brachy podium pinnatum, Moilinia caerulea)*. The detritus decomposes only slowly due to the high cellulose content and moisture stratification with extremes of wet and dry straw. Nutrients are minimal on the surface.

Controlled burning is not appropriate for grassland ecosystems with plants that do not completely die back in winter or with plants having exposed reproductive organs (rosettes, mosses, tuft grasses) or where bulky plants such as legumes and forbs, dying back early, suffocate other species before burning can be of use. These ecosystems include ''Glatthaferwiesen'' (tall oat grass meadows), ''Hochstaudenfluren'' (large forb communities dominated by *Filipendula ulmaria)*, and ''Sumpfdotterblumenwiesen'' (eutrophic wet meadows characterized by *Caltha palustris)*. Green[54] and Schiefer[62] suggest the use of controlled winter burning in combination with other management options. Rotational systems for differing grassland types are optimal for restoration and maintenance of these ecosystems. Ideally, these programs resemble traditional systems of management. With their metamorphosis cycles and their flowering and fruiting cycles, the species associated with grasslands are adapted to specific management cycles (periodic perturbations).

VI. OTHER RESTORATION METHODS

Only a selection of restoration methods with application to nature conservation has been discussed. They are of importance because the affected ecosystems were once large areas of the landscape and have recently become conspicuously rare. Many threatened species find their last refuge in these areas. Other methods include:

1. Installment of pesticide and fertilizer exclusion strips and fields in areas of intensive agriculture (row crop farming)
2. Installment of native grass and forb strips in row crop farming areas and on highway embankments

3.   Management of coppice stools and osiers for basketry
4.   Management of farm ditches
5.   Recreation of meandering streams
6.   Management of charcoal forest, forest pasture, and other forms of ancient forest use

Special techniques are necessary for these restoration methods, of which many do not have a wide applicability. More details can be found in Bruns.[33]

Finally, the techniques of planting trees and shrubs and of applying seed mixtures should be mentioned. Because of the small number of commercially available seed mixtures and because of the uncertainty of their content and place of origin, these techniques have only limited applicability. Planting trees is not the same as planting a forest. The soil and associated life forms have not developed, and most animal species are lacking. The result is deceiving.

VII. CASE STUDIES

Case studies serve two purposes. They demonstrate the application of some of the methods described previously, as well as the use of ecological data as a point of entry into general landscape planning. They also clarify the meaning of nature conservation as a goal of restoration planning. The details of these case studies are given elsewhere.[63,64] Here, the planning strategy is of main interest.

Both case studies result from commissions received by the Institut für Landschaftsplannung (under the direction of Prof. G. Kaule). The following material is a result of combined efforts and is not the work of the author alone.

A. Case Study I — Restoration of the Fliede-Fulda Marsh

A high speed rapid transit railway connection between Hannover and Wurzburg in West Germany is presently under construction. On its way from north to south, the train route passes through the state of Hesse and destroys 4 ha of a 14-ha large area of wetland at the confluence of the Fulda and Fliede rivers. The wetland contains primarily *Carex disticha* and *C. gracilis* marshes, *Phalaris arundinacea* reeds, and herbaceous wet meadows rich in species diversity. There are various ditches and several relicts of a former river channel (ox bow lakes). The fauna of this area includes wetland birds, amphibians, and specially adapted insects. Many of the identified species can be found in the "Red Data" books.

According to federal law, the destruction of parts of this area require compensatory measures. For this purpose, Bruns and Kaule[63] developed the following strategy (c.f. Table 3):

1.   Preservation of the largest area of original wetland possible for the longest period possible, in order to lengthen the time required for species donation to the new developing habitats.
2.   Preparation of substitute habitats before the destruction of parts of the original wetland, so that at least temporarily, the new areas are bordering the intact old wetland habitats.
3.   Installation of pond ecosystems of different nutrient levels, using the methods of regulation of the water regime, implantation of pond reeds, and primary succession of water communities, pond edges, and areas subject to intermittent natural flooding. The incorporation of differing nutrient levels into the design was necessary to provide habitats for all species threatened by the pending changes.
4.   Development of large, intermittently submerged sedge marshes through regulation of the water level and managed secondary succession.
5.   Development of forb communities on irregularly distributed excavation material, mainly topsoil containing propagules, which also serves to control ground and surface water flow.

Table 3
PLANNING PHASE SCHEDULE FOR CASE STUDY I

Date	Railway construction	Habitat restoration
3/83		Fencing of existing wetland area
		Excavation of new ponds
		Implantation of sods
		Removal of top soil (scraping off)
		Connection of existing and new ponds
8/83	Construction of service roads	Transplantation of bushes and trees
		Installment of tongue and groove steel for maintaining the water table
9/83	Excavations for railway bridges	Adjustment of fencing to progress of construction
11/83	Excavations and foundations for retention walls	
1/84	Completion of retention walls, foundations for bridge pillars	Excavation of additional ponds as compensation for unexpected impacts
		Adjustment of fencing
4/84	Beginning of dam construction	Adjustment of fencing to progress of construction
7/84		Final fencing adjustment, permanent habitat area reached
7/85	Completion of bridges	
6/86	Completion of dam	Hydroseeding and planting of dam enbankments

6.　Observation of the development of the different habitats with a monitoring program. Every year an inventory of plants and certain animal species will be conducted. Directives for future problems will be obtained from these results.

The project began in 1983 and the habitats are, at the time of this writing, developing as hoped. Figure 1 describes a simplified schema of the original situation and the resulting habitats.

B. Case Study II — Restoration of a Creek Ecosystem

As a result of land consolidation measures completed in 1980 on a steep vinyard north of Stuttgart, a "creeping landslide" has begun. The vinyard sits on a base of gypsiferous Marl (triassic). In order to save the sliding vinyard, a scheme had to be devised that took into consideration the underground sliding profile. Geological engineers calculated that in order to stop the shift, a large weight (a soil dump, landfill site) had to be installed at the base of the affected hillside. Unfortunately, the mass of earth required will cover a rare ecosystem at the base of the hillside, an oligotrophic creek located in triassic sandstone and bordering swamp forests. In addition, neighboring old orchards, a site of 10 to 15 years of secondary succession, will be destroyed. The species diversity of this unit is extremely high and includes rare birds, insects, plants, and an extremely rare crustacean *Astacus torrentinum*. Many of these species are listed in our "Red Data" books.

According to the law, compensatory measures for the destruction of this habitat were required. For major sections of the creek, no compensation is possible as discussed earlier. The following strategy was developed:

1.　Preservation of areas for which no restoration of the ecosystem is possible, i.e., crustacean habitats and natural riparian swamp forest. Changes in the engineering design can reduce the extent of landfill through retention gabiones (wire structures filled with rock).

2.　Completion of substitute habitats before the destruction of the original creek.

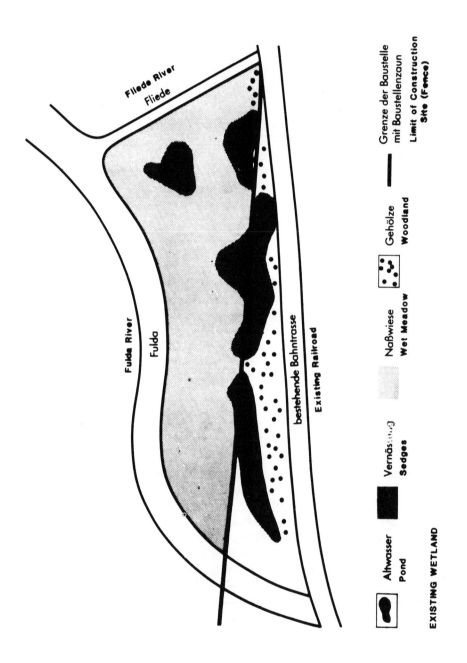

Fliede River
Fliede

Fulda River
Fulda

bestehende Bahntrasse
Existing Railroad

Grenze der Baustelle
mit Baustellenzaun

Limit of Construction
Site (Fence)

Gehölze
Woodland

Naßwiese
Wet Meadow

Vernässung
Sedges

Altwasser
Pond

EXISTING WETLAND

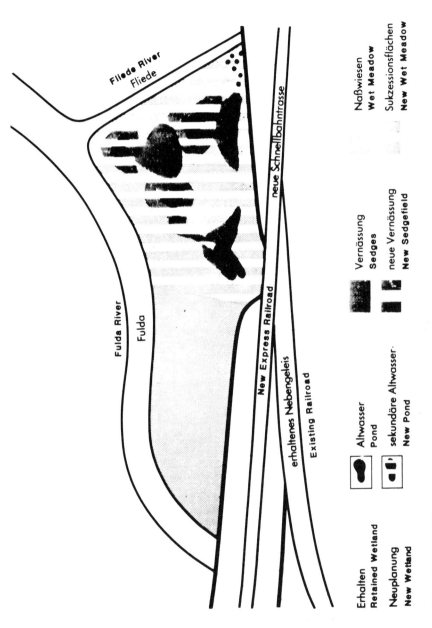

CREATED WETLAND

FIGURE 1.   Case study I, restoration of the Fliede-Fulda marsh, simplified schema of the original situation (top) and the resulting habitats (bottom).

3. Transplantation of small sedge marsh pockets from the riparian swamp forest to the newly prepared areas to provide for new wetlands and a new water course.
4. Development of a new stretch of creek bed between the preserved sections of the species rich creek with the help of natural colonization of gravel beds; preparation of a wide bed in which the creek can find its own course on the existing base of sandstone, gravel and rocks; by implantation of sedge sods and transfer of original stream bed rocks; by plantation of alders and willows for shading and stabilization of the banks.
5. Promotion of grassland ecosystems on the new landfill site through natural colonization as a substitute for existing areas with open Marl and pioneer vegetation, and through secondary succession and implantation using existing top soil obtained by clearing of the landfill site.
6. Planting of fruit trees and alders as a replacement for the lost orchards and riparian forest.
7. An ongoing monitoring program as described in Case Study I.

The new creek is not identical to the old one. It is as if an element of the European postglaciation period were placed into the cultural landscape. It is only a substitute; however, one with a chance to evolve through undisturbed succession into a regionally typical riparian ecosystem. Figure 2 describes a simplified schema of the original situation and the resulting habitats.

C. General Observations

During construction, the preservable relics and the new ecosystems must be kept strictly separated from the construction sites (rail construction, landfill clearing in the case studies). A 2-m high fence can be used. Contaminated, effluent rich water from the construction site must be channeled through a separate drainage system. In Case Study II, no earth can come close to the creek or swamp forest. Hillside drainage above this area must be redirected. For this purpose, a retention wall was designed for the foot of the hillside to be built before preparation of the landfill site. The wall will be planted and has a fence.

Compensation measures often require a reorganization of the original ecosystem into a new scheme of ecosystem elements with reference to time and space (Figure 2). Structurally complex units, often with a history of several decades of succession, will be taken apart, and the components will be used to design new habitats. In order to achieve functional units equivalent to the original habitats the new areas must usually be larger than the old ones. In Case Study I, the new wetland area had to be larger by about 20% in order to provide both sufficient space for some especially important and sensitive species, and various stages of succession. The area calculation was done on the basis of the territorial needs of these species in consideration of their new distribution.

It is extremely important for success that the restoration measures be carried out as neatly and technically accurate as possible. The value of the physical presence of the responsible planner on the construction site cannot be underestimated. Details are often decisive. For example, in Case Study I, the preservation of ditches containing sedges was necessary for the development of sedge marshes on raw alluvial soils. The excavator must be accordingly directed. On other sites, in order to allow for a variety of environmental conditions on a small scale, it was important *not* to smooth out the embankments but to leave a rough surface. These are on-the-spot decisions. The diversity of the alluvial soils produces surprises. During railway construction, underground peat appeared instead of gravel. The foundation for the dam had to be extended, which required a quick change of plans regarding the location and size of the new habitats.

Working with succession only works when neighboring species pools exist. Otherwise, undesirable species will invade.

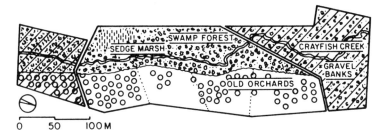

Original situation of proposed land fill site with areas to be preserved shaded

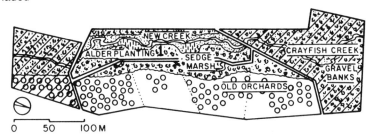

Phase 1: storage of sedge marsh sods, construction of soil retention and drainage, preparation of new creek bed, implanting sedge pockets and rocks

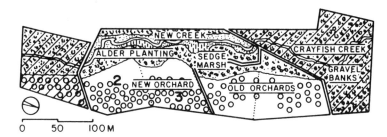

Phases 2 to 5: dumping of soil during the 4 years following phase 1, alder planting and covering of land fill with marl after completion of each phase

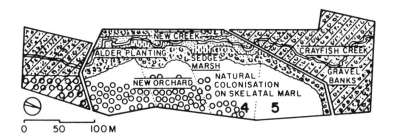

Phase 6: completion of alder strips, planting of new orchard, monitoring ecosystem development

FIGURE 2.   Case study II, restoration of a creek ecosystem, simplified schema of restoration strategy (for details, see text).

VIII. DISCUSSION

The great loss of native species in Europe is related to the loss of environmental extremes. Average conditions have become the rule. The nutrient load increased exponentially in all regions through heavy fertilization. The eutrophication of the landscape is the final result. Poorly protected habitats and sites of special scientific interest are sandwiched between urban conglomerates and fields of intensive crop production. Most conservation sites are too small to prevent fertilization by wind drift and surface runoff from adjacent sites, herbicide and pesticide impacts, changes in groundwater, and visitor impact.[6,11]

The problems resulting from lack of large conservation areas was recognized early. Seifert,[65] in 1936, called for designation of at least 10% of the cultural landscape as nature preserves. Recent assessments give more differentiated data. Kaule[18] asks for a regionally adapted conservation strategy with about 5% of the region needing full protection, nature conservation being the only land use. For a network of linear elements and small habitats such as ponds, another 5% or more (up to 35%) is needed to connect the preserves and serve as stepping stones for species dispersal. The 15 to 30% of the commercially used land will need to be operated according to special guidelines, in order to preserve species adapted to special farming or forestry techniques. Presently, the fully protected area is less than 1% of national land.

It is not sufficient for a national conservation strategy to limit itself to maintaining and protecting existing preserves. Restoration of damaged ecosystems to meet legal requirements is not enough. In the present critical situation, there is an inherent need for the creation and reconstruction of ecosystems for their own sake. Here, ecological qualities that have largely disappeared from the cultural landscape must have priority. The association of most extinct and threatened species with ecosystems of extreme ecological qualities (such as extremes of moisture, harsh, infertile soil, and exposed and arid sites) was discussed. These ecosystems may result from periodic perturbations. Perturbations are important in maintaining and restoring natural ecosystems,[66-68] but obviously anthropogenic ecosystems are just as perturbation dependent, if not more so. In the cultural landscape, perturbation caused by traditional land management predominate over natural perturbation. Traditional systems of management include crop rotation, historically with cycles lasting up to 12 years, and rangeland management where areas were regularly overgrazed and then abandoned to recover.[54]

For successful restoration efforts, it will be necessary to learn more about the role of perturbation in ecosystem development. Thus, it should be possible to restore just those ecosystems which are, at the present time, most rare and endangered. Ecosystems that have nutrient poor stages at the beginning of their evolution lend themselves to "implantation" and "natural colonization". However, long periods of succession must be accepted, which usually exceed the time period allowed for legal compensation. For ecosystems dependent on frequent perturbation, the goal of restoring "low fertility" conditions can usually be achieved within the liability period by the application of special management regimes.

The problem of the next few years will be whether it is possible with help of present and evolving restoration methods to develop as much land as is necessary for nature conservation (Table 4). This is not only dependent on the static accumulation of land, but requires an ongoing strategy of renewal. Primary succession, for example, is only possible if species suitable for colonization are nearby, and if skeletal soils are available. This means that the mining industry and nature conservation must devise a common long-range strategy, and other groups interested in the use of abandoned surface mines must relinquish their claims.[69,70]

The task of the ecological sciences is to give additional support to the demands for more areas of the cultural landscape for nature conservation purposes. It is also of considerable importance to determine what minimal density of different habitat types is required to insure long-term stability and renewal of ecosystems.

Table 4
SUMMARY OF EXAMPLES OF ECOSYSTEM RESTORATION AND
ECOLOGICAL QUALITIES RESTORED

Ecosystem	Restoration method	Restored quality
Abandoned peat cutting	Regulation of water regime	Low fertility
	Natural colonization	High moisture
Oligotrophic fen community	Natural colonization of gravel pits	Low fertility
		High moisture
Oligotrophic lakes	Natural colonization of gravel pits	Low fertility
		High moisture
Wetland with ponds	Implantation, regulation of water regime	Medium fertility
		High moisture
Sedge marsh	Regulation of water regime	Medium fertility
	Implantation	High moisture
Seminatural streams (sections)	Recreation of meanders, providing new stream bed	Medium fertility
		High moisture
	Natural colonization	
Carr, swamp forest	Regulation of water regime	High fertility
	Natural colonization	High moisture
Reeds	Natural colonization	High fertility
	Implantation	High moisture
Heath	Winter burning (extraction of biomass)	Low fertility
		High exposure
Gravel banks and islands	Regulation of water regime	Low fertility
	Natural colonization	High exposure
Dry chalk grassland	Extraction of biomass (mowing)	Low fertility
	Natural colonization	High exposure
Hedgerows, road embankments	Coppicing, transplantation, skeletal soil infill	Medium fertility
		Medium exposure
Tall-grass meadows	Extraction of biomass (mowing)	Medium fertility
	Natural colonization	Medium exposure
Forb communities	Secondary succession	High fertility
	Periodically managed	Medium exposure

IX. CONCLUSIONS

Dramatic changes in the traditional Central European landscape and the catastrophic rate of extinction of natural flora and fauna during the last few decades require special attention when considering management options for restoration and enhancement of ecosystems. A continuing loss of wildlife habitats will, in the near future, lead to the extinction of more than 50% of our indigenous species. Hence, the organized public interest in conservation is large and public pressure on political decisions has resulted in new laws that direct future land management. Ecosystem restoration must serve the purposes of conservation. My discussion illustrates, through a review of pertinent literature, several forms of ecosystem management, including ecosystem-transplantation and creation of new ecosystems. Two recent restoration projects were summarized, where the emphasis is on the problems caused by excessive fertilization and habitat destruction in the cultural landscape.

ACKNOWLEDGMENTS

I would like to thank my wife Nancy for her valuable assistance in the preparation of this paper. I also thank Dr. B. H. Green, Dr. R. S. Dorney, and Prof. Dr. Kaule for their detailed comments.

REFERENCES

1. **Bradshaw, A. D. and Chadwick, M. J.,** *The Restoration of Land,* Blackwell Scientific, Oxford, 1980.
2. **Bauer, H. J.,** Ten years studies of biocenological succession in the excavated mines of the Cologne lignite district, in *Ecology and Reclamation of Devastated Land,* Vol. 1, Hutnik, R. and Davis, G., Eds., Gordon & Breach, New York, 1973, 271.
3. **Schiechtl, H.,** *Bioengineering for Land Reclamation and Conservation,* University of Alberta Press, Edmonton, 1980.
4. **Leonardson, L. and Ripl, W.,** Control of undesirable algae and induction of algal succession in hypertrophic lake ecosystems, *Developments in Hydrobiology,* Vol. 2, 1980.
5. **Klötzli, F.,** Some aspects of conservation in overcultivated areas of the Swiss Midlands, in *Wetlands: Ecology and Management, Proc. 1st Int. Wetlands Conf.,* New Delhi, India, September 1980, Vol. 2, first published as *Int. J. Ecol. Environ. Sci.,* Vol. 7, 1981.
6. **Green, B.,** *Countyside Conservation,* Allen & Unwin, London, 1981.
7. **Thomas, W. L., Ed.,** *Man's Role in Changing the Face of the Earth,* University of Chicago Press, Chicago, 1956, 183.
8. **Kelcey, J. G.** Industrial development and the conservation of vascular plants, with special reference to Britain, *Environ. Conserv.,* 11(3), 1984.
9. **Klötzli, F.,** Zur Frage der Neuschaffung von Mangelbiotopen, in *Gefährdete Vegetation und ihre Erhaltung,* Ber. Int. Symp. Internationalen Vereinigung für Vegetationskunde, J. Cramer, Vaduz, 1981.
10. **Sukopp, H.,** Bewertung und Auswahl von Naturschutzgebieten, *Schriftenr. Landschaftspflege Naturschutz,* 6, 183, 1971.
11. **Heydemann, B.,** Die Bedeutung von Tier- und Pflanzenarten in Ökosystemen, ihre Gefährdung und ihr Schutz, *Jahrb. Naturschutz Landschaftspflege,* 30, 15, 1980.
12. **Blab, J., Nowak, E., Trautmann, W., and Sukopp, H.,** *Rote Liste der gefährdeten Tiere und Pflanzen in der Bundesrepublik Deutschland,* Kilda Verlang, Greven, 1984. (Official Red Data Books for the FRG).
13. **Sukopp, H., Trautmann, W., and Kornek, D.,** Auswertung der Roten Liste gefährdeter Farn- und Blütenpflanzen in der Bundesrepublik Deutschland für den Arten- und Biotopschutz, *Schriftenr. Vegetationsk.,* 12, 1978 (incl. English summary).
14. **Jürging, P. and Kaule, G.,** Biotopkartierung für die Landschaftsrahmenplanung, *Schriftenr. Naturschutz Landschaftspflege,* 8, 7, 1977.
15. The Federal Nature Protection Act: "Bundesnaturschutzgesetz: of 1976, Federal law gazette I, p. 3574, amended 1977 and 1980. For quick reference see: *Functions and Organisations for Environmental Protection in the Federal Republic of Germany,* Umweltbundesamt, Ed., Berlin, 1984.
16. **Bruns, D. and Hoppenstedt, A.,** The environmental impact assessment, an important new planning tool for Ireland, presented at Int. Semin. Environmental Diplomacy: The Management and Resolution of Transfrontier Environmental Problems, Ennis, County Clare, Ireland, 1985.
17. **Kaule, G. and Schober, M.,** Möglichkeiten und Grenzen des Ausgleichs von Eingriffen in Natur und Landschaft, *Schriftenr. Bundesminist. Ernährung, Landwirtschaft und Forsten: Angew. Wiss.,* 314, 1985.
18. **Kaule, G.,** *Arten- und Biotopschutz,* Universitätstaschenbücher Große Serie, Ulmer Verlag, Stuttgart, 1986, in press.
19. **Williams, R.,** Floral world's fair, *Landscape Archit.,* 408, 1980.
20. **Ringler, A.,** A third hand biotope, *Garten Landschaft.,* 465, 1983.
21. **Pollard, E., Hooper, M. M., and Moore, N. W.,** *Hedges,* The New Naturalist, London, 1977.
22. **Reschke, K.,** Lebende Hecken werden versetzt, Neue Arbeitsweisen in der Flurbereinigung, *Natur und Landschaft,* 55, 351, 1980.
23. **Unger, H.-J.,** Verpflanzung von Hecken und Feldrainen im Rahmen der Flurbereinigung, *Natur und Landschaft,* 56, 295, 1981.
24. **Klötzli, F.,** Naturschutz im Flughafengebiet: Konflikt und Symbiose, *Flughafen-Information,* Vol. 3, Zürich, 1975, 3.
25. **Burnard, J.,** Erhaltung von Streu- und Moorwiesen durch Verpflanzung, in *Gefährdete Vegetation und ihre Erhaltung,* Ber. Int. Symp. Internationalen Vereinigung für Vegetationskunde, J. Cramer, Vaduz, 1981.
26. **Klötzli, F.,** Zur Verpflanzung von Steu- und Moorwiesen, *Tagungsber. Akad. Naturschutz Laufen,* 5, 41, 1980.
27. **Odum, E. P.,** *Fundamentals of Ecology,* W. B. Saunders, Philadelphia, 1971.
28. **Mayser, W.,** Rhein-Main-Donau AG, Regensburg, 1985, personal communication.
29. **Bradshaw, A. D.,** The reconstruction of ecosystems, *J. Appl. Ecol.,* 20, 1, 1983.
30. **Lee, J. A. and Greenwood, B.,** The colonisation by plants of calcareous wastes from the salt and alkali industry in Cheshire, England, *Biol. Conserv.,* 10, 131, 1976.

31. **Andreae, M. I. and Cavers, P. B.,** The significance of natural vegetation in abandoned gravel pits, in *Revegetation of Pits and Quarries,* Suffling, R., Ed., School of Urban and Regional Planning, University of Waterloo, Ontario, Working Paper No. 13, 1981.

32. **Pfeiffer, H.,** Vom gesetzlichen Verhalten der Pioniere bei Neulandbesiedelung, *Mitt. Floristisch-Soziol. Arbeitsgem.,* 10, 87, 1963.

33. **Bruns, D.,** Planung von Ersatzbiotopen, Ph.D. thesis, Universität Stuttgart, West Germany, 1986.

34. **Beutler, A., Kaule, G., Scholl, G., Schwenninger, H., and Seidl, F.,** Ökologische Wirkungen unterschiedlicher Wirtschaftswegetypen, Research report commissioned by Landesamt für Flurbereinigung und Siedlung Baden-Württemberg, 1984.

35. **Davis, B. N., Ed.,** *Ecology of Quarries,* Institute of Terrestrial Ecology, ITE Symposium, Vol. 11, 1982.

36. **Wartner, H.,** *Steinbruüche vom Menschen geschaffene Lebensräume,* Landschaftsökologie Weihenstephan, Vol. 4, 1983.

37. **Dahl, H. J., Jürging, P.,** Abgrabungen als Sukzessionsfläche für Flora und Fauna, *Jahrb. Naturschutz Landschaftspflege,* 32, 55, 1982.

38. **Heydemann, B.,** Die Bedeutung der Kiesgruben als Renaturierungsgebiete, *Jahrb. Naturschutz Landschaftspflege,* 32, 93, 1982.

39. **Kelcey, J.,** *Ecological Studies in Milton Keynes: Brickfields (4),* Milton Keynes Development Corporation, 1974.

40. **Mahler, U., Röben, P., and Vogt, D.,** Zufluchtsinseln für bedrohte Tier- und Pflanzenarten, *Jahrb. Vereins zum Schutze der Bergwelt,* 45, 135, 1980.

41. **Zielonkowski, W.,** Wildgrasfluren der Umgebung Regensburgs, Ph.D. thesis, Universität München, West Germany, 1973.

42. **Ash, H. J.,** The Natural Colonisation of Derelict Industrial Land and its Development for Amenity Use, Ph.D. thesis, University of Liverpool, England, 1983.

43. **Jürging, P. and Kaule, G.,** Entwicklung von Kiesbaggerungen zu ökologischen Ausgleichsflächen, *Naturschutz Landschaftspflege,* Bayerisches Landesamt für Umweltschutz, München, 1977.

44. **Eigner, J. and Schmatzler, E.,** *Bedeutung, Schutz und Regeneration von Hochmooren,* Naturschutz aktuell, 4, 1980.

45. **Schwaar, J.,** Möglichkeiten und Grenzen der Moorregeneration. Erfahrungen in Nordwestdeutschland, Tagungsberichte, Vol. 6, Akademie für Naturschutz und Landschaftspflege, Laufen/S., 1981.

46. **Tscharntke, T.,** Klärteiche: Feuchtgebiete in einer ausgeräumten Kulturlandschaft, *Natur und Landschaft,* 58, 333, 1983.

47. **Reichholf, J. and Reichholf-Riem, H.,** Die Stauseen am unteren Inn: Ergebnisse einer Ökosystemstudie, *Ber. Bayer. Akad. Naturschutz und Landschaftspflege,* 6, 1982.

48. **Grünig, A.,** Die Vegetationsentwicklung im Flachseegebiet, *Jahrbuch Stiftung Reusstal,* ETH Zürich, 1977, 16.

49. Bayerisches Landesamt für Wasserwirtschaft, Ökotechnische Modelluntersuchung Untere Isar, Research report, München, 1983.

50. **Catchpole, C. K. and Tydeman, C. F.,** Gravel pits as new wetland habitats for the conservation of breeding bird communities, *Biol. Conserv.,* 8, 47, 1975.

51. **Niemann, E. and Wegener, U.,** Verminderung der Stickstoff- und Phosphoreinträge in wasserwirtschaftliche Speicher mit Hilfe nitrophiler Uferstauden und Verlandungsvegetation: ''Nitrophyten-Methode'', *Acta Hydrochim. Hydrobiol.,* 4, 269, 1976.

52. **Harrison, J.,** *The Sevenoaks Gravel Pit Reserve,* WAGBI Publication, Chester, England, 1974.

53. **Dingethal, F. J., Jürging, P., Kaule, G., and Weinzierl, W.,** *Kiesgrube und Landschaft,* Parey Verlang, Hamburg, 1985.

54. **Green, B. H.,** The management of herbaceous vegetation for wildlife conservation, in *Management of Vegetation,* Way, J. M., Ed., British Crop Protection Council, Monogr. No. 26, Croydon, 1984, 99.

55. **Ellenberg, H.,** *Die Vegetation Mitteleuropas mit den Alpen,* Ulmer Verlag, Stuttgart, 1978 (New edition, 1985).

56. **Rodi, D.,** Maßnahmen zur Verhinderung der Verbuschung der Trockenrasenstandorte des Naturschutzgebietes ''Bargauer Horn'' bei Schwäbisch Gmünd, in *Ber. Int. Symp. Internationalen Vereinigung für Vegetationskunde,* J. Cramer, Vaduz, 1981.

57. **Schiefer, J.,** Ergebnisse der Landschaftspflegeversuche in Baden-Württemberg, *Natur und Landschaft,* 58, 295, 1983.

58 **Schiefer, J.,** Möglichkeiten der Aushagerung von nährstoffreichen Grünflächen, *Veröff. Naturschutz Landschaftspflege Baden-Württemberg.* 57/58, 1984.

59. **Krause, W.,** Ausbreitungsfähigkeit der Niedrigen Segge *(Carex humilis), Planta,* 31, 91, 1940.

60. **Salisbury, E. J.,** *The Reproductive Capacity of Plants,* Bell and Sons Ltd., 1942.

61. **Riess, W.,** Der Feuereinsatz und seine Technik in der Landespflege, *Natur und Landschaft,* 51, 284, 1976.

62. **Schiefer, J.,** Kontrolliertes Brennen als Landschaftspflegemaßnahme?, *Natur und Landschaft,* 57, 264, 1982.

63. **Bruns, D. and Kaule, G.,** DB-Neubaustrecke Hannover-Würzburg PA. 17.4 (Fulda-Fliede-Aue), Gutachten über Eingriffe in den Naturhaushalt und Konzepte für Ersatzmaßnahmen, Ecological study commissioned by the German Federal Railway Authority, Frankfurt, 1983.

64. **Bruns, D. and Kaule, G.,** Hangrutschung Sommerberg (Erlenbach bei Heilbronn), Landschaftsökologischer Beitrag zur Sanierung, Ecological study commissioned by the State Authority on Land Consolidation, 1985.

65. **Seifert, W.,** Die Versteppung Deutschlands, *Dtsch. Tech.,* 4, 423, 1936.

66. **Knapp, R. Ed.,** *Vegetation Dynamics,* 1974.

67. **Connell, J. H. and Slatyer, R. O.,** Mechanisms of succession in natural communities and their role in community stability and organisation, *Am. Nat.,* 111, 1119, 1977.

68. **Vogl, R. J.,** The ecological factors that produce perturbation-dependent ecosystems, in *The Recovery Process in Damaged Ecosystems,* Cairns, C., Ed., Ann Arbor Science, Ann Arbor, Mich., 1980, 63.

69. **Dorney, R. S.,** Reclamation: Sometimes the 'Cure' is worse than the 'Disease', *Landscape Archit.* 74, 1984.

70. **Kaus, D.,** Open-cast workings. An opportunity for species conservation and the stabilization of the natural equilibrium, *Garten Landschaft.,* 172, 1979.

INDEX

A

Abandoned hazardous waste sites, 62
Abandoned mines
  erosion control, 54—55
  fertilizer, 55
  Longwall Mine District (Illinois case history), 40,
      45—49, 54, see also Illinois Longwall Mine
      District
  moisture conservation, 54—55
  organic material, 55
  pH alteration, 55
  plant selection, 55—56
  reclamation suggestions, 54—56, see also other
      subtopics hereunder
  revegetation problems, 40—44, see also Revegeta-
      tion
  salt management, 55
  spoil materials, 44, 47
  western surface mines (North Dakota case history),
      40, 49—54, see also North Dakota surface
      mines
Acceleration of succession, 17
Acid-base accounting, 145
Acidic soils, 42
Acidity, 42, 55
Acid test, 14, 19—20
Aerial colonization, 30—31
Aerial inventory methods wetlands, 120—121
Agricultural research, 14
Agro-ecosystems, 16
A1, 47—49
Alternative ecosystems, 10
Applied ecology, 2, 6, 13—14
Approximate original contour (AOC), 140—141,
      143, 150, 156—160
  alternative to, 152—155
  backfill stability, 151—152
  controlled overburden placement procedures, 151
  reclamation to, 151
  variance, 151, 159
Aquatic insects, 30
Aquatic invertebrates, 31, 32
Architecture of species, 19
Artificial boulders, 35
Artificial coral reef/lagoon microcosm, 17
Artificial disturbances, 19
Artificial forest communities, 17—18
Artificial wetlands, 8
Assessment of success or failure, 130—132
Atmospheric pollution, 66
Automatic monitoring, 84
Ayrshire Mine Slurry Project, 93—97, 102—108, see
      also Coal slurry

B

Backfill construction, 151—152

Bank covers, 33
Barren sites, 56
Barrier system, 79
Bermuda Biological Station, 3
Biological desert, 164
Biological elements, 3, 5, 118—120
Biomass removal, 173—175
Biotechnology, 2
Bottomland hardwoods, 118
Boulder placements, 31, 33—35
Brackish marsh, 128
Buffer strip, 28, 33
Building on contaminated land, 84—87

C

Ca, 47, 49, 52
California salt marsh restoration, 123—138, see also
      Salt marsh
Capping, 78
Carbon dioxide, 82
Central Appalachian mining region, 143—144
CERCLA, 62
Channel geometry, 25
Chemical treatment of soil, 73
Climatic diagram, 46, 51
Coal-mine lakes, 116
Coal slurry acid, 92—93, 96, 98—100
  Ayrshire Mine Slurry Project, 93—97, 102—108
  characterization, 92—93
  composition, 92
  creation of wetland habitat in, 91—114, see also
      other subtopics hereunder
  cyclone separation, 98—99, 108—110
  defined, 92
  discharge, 92—93, 97, 99—100, 111—114
  disposal ponds, 92
  hydrologic and hydraulic considerations, 100—101
  Leahy Mine cyclone project, 98—99, 110—111
  leap frog approach, 98—99
  management techniques, 97—100
  mechanical manipulation, 97
  neutralization potential, 93, 96, 99
  pH, 96
  preexisting basins, 92—97
  revegetation, 95—96, 99—101
  seed mixture applied, 94
  soil moisture, 92
  Sun Spot Mine project, 100
  treatment, 93, 98—99
  tree plantings, 94—95
  vegetation, 95
  wetland plants, 93—95
  wildlife, 96—97
  zonation, 93, 96
Coastal wetlands, 124
Colonization pattern, 18, 30, 32
Combustible materials, 87

Compaction, 44, 146
Containment, 76—82
Contaminants, 78, 80, 83, 86
Contaminated land
  building on, 84—87, see also Building on
    contaminated land
  containment, 76—82
  covering systems, 78—82, see also Covering
    systems
  defined, 62
  difficulties of excavation, 66, 68—69
  hazards/exposure combinations, 64—65
  in-ground barriers, 77—78
  in situ treatment, 69, 75—76
  isolation, 76—82
  macroencapsulation, 69, 76—82
  on-site treatment, 69
  physical hazards, 65
  reclamation and treatment of, 61—89, see also
    other subtopics hereunder
  recycling, 62, 66
  remedial measures, 66—71, 82—84
  removal of contamination, 66—67
  risk assessment, 64
  soil treatment after excavation, 71—74, see also
    Soil treatment
  sources of contamination, 63—64, 67
  targets at risk, 64—65
  temporary amelioration, 66
  toxic hazards, 65
  treatment methods, 69—70
  urban environment, 65—66
Contamination, 63—64, 67, 69, 81—82
Controlled overburden placement, 143—147
Convex stream bed, 33
Cordgrass, 126—131, 133
Covering system, 33, 78—82, 84
Creek ecosystem, restoration of, 177, 180—181
Creeping landslide, 177
Critical parameters, 16, 19—20
Cu, 52
Current deflectors, 33—34
Cyclone separation, 98—99, 108—110

D

Dams, 33—34
Daucus carota, 19
Deep-mined areas, 40, see also Abandoned mines
Deflectors, 31, 33—34
Depth characteristics, 31
Design life, 71, 83
Dilution of substances, 28
Dipsacus sylvestris, 19
Disruption, 16
Disturbed ecosystems, 3
Diversity, 126—127
Domestic refuse, 82

Double wing deflectors, 31, 34
Drainage installation, 83, 84
Drift of organisms, 30—31, 33, 49—50

E

EC, 47, 52—53
Ecological discovery, pattern in, 15
Ecological research, 13—15
Ecological significance of reactions, 17
Ecological theory and ecological practice, relation-
  ship between, 20—21
Economic development, 153
Ecotechnology, 124, 129—130
Effectiveness of remedial measures, 70—71
Electrical conductance, 43
Empirical experiments, 19, 26
Endangered species, 126, 128, 131, 133—134, 166
Endangered Species Act, 124
Environmental factors, 40
Equilibrium, 25—26, 31
Erosion control, 28, 44, 49, 54—55
ESP, see Exchangeable sodium percentage
Ethical concerns, 21
Evaluation of performance, 82—84
Excavation, 66, 68—69, 71—74, 84
Excess spoil disposal, 150, 156
Exchangeable sodium percentage (ESP)), 43
Experimental ecology, paradigm for, 16
Experimentation in salt marshes, 129—130
Extinction, 124, 165

F

Fe, 47, 52—53
Federal conservation law, 166
Fertility, 42—43, 49
Fertilization, 42, 55
Fish habitat enhancement, 33—35, 118
Flexible membranes, 80
Fliede-Fulda marsh, 176—179
Flooded gravel pits, 172—173
Fritted subsoils, 29
Functional aspects of ecosystem, 3, 6
Fusion/vitrification, 76
Future needs, 10—11

G

Gabion deflectors, 31, 34
Gas, 80, 82, 86
Genetically altered organisms, 2
Geologic factors, 25
Geologic materials as topsoil substitutes, 144—145
Geomorphic features, 24
Geotextiles, 80
Ground engineering, 85—86
Groundwater, 69, 70, 77, 78, 81, 119

Gypsum, 55

H

Habitat characteristics, 31
Hazardous waste sites, 2, 8—10, 62
Hazards/exposure combination, 64—65
Heat treatment of soil, 74
Heuristic value, 13, 19
Highwall backfills, 151
Hydraulic measures, 77
Hydrological planning, 24—27, 127—130, 135

I

Illinois Longwall Mine District, 40, 45—49
  comparison with North Dakota sites, 54
  general site description, 45
  Ladd site, 41, 48
  site assessment, 49
  Spring Valley site, 41, 46—48
  standard site, 41, 48
  study sites, 41, 45—49
  Wenona site, 41, 48—49
Imitation of natural systems, 20
Implantation, 169—170
Indiana wetlands, 115—122, see also Wetlands
Indoor air quality, 66
Industrialization, 66
In-ground barriers, 77—78, 83—84
In situ soil treatment, 69, 75—76, 83
Institutional constraints to improved reclamation,
  158—160
Instream habitat structures, 31, 34
Irrigation, 29, 42, 55
Isolation technique, 27—28, 76—82

K

K, 47, 49, 52—53

L

Lake eutrophication, 17
Leaching, 43, 55, 69
Leahy Mine Cyclone project, 98—99, 110—111
Life-history traits, 18
Lime requirements, 47—49
Littoral zone, 120
Log-drop structures, 31
Long-term monitoring program, 83
Lotic environments, 24

M

Macroencapsulation, 69, 76—82
Macroinvertebrates, 30—33
Management goals, 7
Massive subsoils, 29
Meander parameters, 25—26
Mesic, 17—18

Methane, 82, 86
Mg, 47, 52—53
Microbial treatment of soil, 73—74
Micronutrients, 55
Mines, see Abandoned mines
Mine soils, 143, 145—147
Mine spoils, 143, 146
Misplacement of plants, 17
Mitigation, 124—125, 129—130, 134
Mobile soil clean-up installations, 74
Moisture conservation, 49, 50, 54—55, 81—82
Monitoring performance, 82—84
Monitoring program, 131—132
Monitoring reclamation, 35—36
Mountaintop removal-valley fill, 151
Mulch, 44, 55
Mycorrhizae, 16—17, 55

N

N, 16, 17, 47—49, 51—53
Na, 52—54
Native plant communities, 126
Native vegetation on minesites in semiarid West,
  restoration of, 16
Natural colonization, 170—171
Natural ecosystems, 63
Natural revegetation, 40, 49—50, 52
Natural succession, 116
Natural topsoils, 143—144
Natural wetlands, 7
N-fixation, 149—150
Nonpoint source pollution, 27
North Dakota surface mines, 40, 49—54
  comparison with Illinois sites, 54
  Fritz site, 41, 51, 53
  general site description, 49—50
  New Salem site, 41, 53
  site assessment, 53—54
  study sites, 41, 50—53
  Velva site, 41, 50—52
Nutrients, 16, 17, 42, 43, 47, 52, 54—55, 73

O

On-site treatment, 69
Organic matter, 48, 52, 54—55
Osmotic stress, 43
Overall mining cost, 154
Overcultivation, 164
Overhangs, 35
Overpour structures, 33

P

P, 47, 51—53
Paradigm for ecological research, 13—16
Partial transplantation, 169—170
Pattern in ecological discovery, 15
Performance standards, 140
Permeability, 33

Perturbed systems, 20
P-fixation, 148—150
pH, 42, 47—49, 51—55, 96
Physical hazards, 65
Physical treatment of soils, 73
Pickleweed, 126—128, 130—131, 133, 135
Planning ecological research, 15
Planning river and stream restoration, 35—36
Plant selection, 55—56
Point source pollution, 27
Polluted land, 62
Ponds for wetlands, 115—122
Pool control area, 31
Postmined soils, 29
Postmining landforms, 150—158
  alternative, 155—156
  approximate original contour, 151, see also
      Approximate original contour
  comparative economics of alternatives, 155—158
  overall mining cost, 154
  reclamation cost, 154—156
  remining, 158
  Powell River Project, 140, 161
Postmining land use, 151
Predictive capabilities of ecology, 6, 21
Prescriptive regulations, 10
Preservation of species, 166
Productivity, 17
Pyrites, 145

R

Radius of curvature, 26
Rapid assessment technique, 35
Rare plant species, 126, 165
Reciprocal transplants, 18
Reclaimed soil, 16, 29, 54—56, 158—160
Reclamation cost, 154—156
Reclamation laws, 40, 45
Reconstructed soil, 29
Recycling of land, 62, 66
Refuse, 87
Regulations, 160
Remedial measures, contaminated land, 66—71,
      82—84, see also Contaminated land
Remining, 158
Removal technique, 28
Research, 130—132
Restoration, 8—10
  definition, 3, 13
  examples, 2—3
  financial needs, 4—5
  future needs, 10—11
  native vegetation on minesites in semiarid West, 16
  new frontier, 1—11
  opposition to developments in, 7
  resistance to involvement in, 6
  symposium, 14
  synthetic approach to ecological research, 13—21
  technique for basic research, 13
  theoretical benefits of, 3—6

  tool for discrimination, 18
  West Germany ecosystems, 163—186
  wetlands, 116
Revegetation, 27, 29
  abandoned mines, 40—44, see also other subtopics
      hereunder
  acidity, 42
  coal slurry, 95—96, 99—101
  factors critical to success, 148—150
  fertility, 42—43
  potential acidity, 42
  salinity, 43
  sodicity, 43
  soil atmosphere, 43—44
  soil texture, 44
  surface mining, 140
  temperature, 44
  topography, 44
  toxicity, 42—43
  water, 40—42
Riparian vegetarian, 27—30
Rip-rap, 35
Risk assessment, 64
River and stream restoration, 23—38
  bank and bed stability, 24
  buffer strip, 28
  carbon copy technique, 26
  channel geometry, 24, 25
  dilution of substances, 28
  empirical approach, 26
  equilibrium, 25—26, 31
  fish and food source, 24
  fish habitat enhancement, 33—35
  geologic factors, 25
  holistic approach, 28
  hydrologic considerations, 24—27
  isolation technique, 27—28
  macroinvertebrates, 30—33
  meander parameters, 25—26
  monitoring, 35—36
  natural approach, 26
  need for, 24
  planning, 35—36
  pollutant loads, 24
  radius of curvature, 26
  rapid assessment technique, 35
  removal technique, 28
  riparian vegetation, 27—30
  sediment transport, 25
  sinuosity, 26
  soil type, 25
  straight ditch channelization, 25
  systems approach, 27
  transfer of a pollutant, 28
  water quality, 27—28
Rocky Mountain Biological Laboratory, 3

S

Safety margins, 83
Salinity, 43, 48—49, 51, 53—55

Salt management, 55
Salt marsh, 123—138
   assessment of restoration success or failure, 130—132
   diversity, 126—127
   endangered species, 126, 128, 131, 133—134
   experimentation, 129—130
   hydrological planning, 127—129, 135
   mitigation, 124—125, 129—130, 134
   monitoring program, 131—132
   native plant communities, 126
   objectives of restoration, 127
   rare plant species, 126
   regional coordination, 125
   research, 130—132
   restoration goals, 124—127
   sample of projects in Southern California, 133—134
   species requirements, 125—126
SAR, see Sodium absorption ratio
Sediment transport, 25
Settling basins, 28
Si, 47
Siltation, 33
Single deflectors, 31
Sinuosity, 26
Siting criteria, 34
Slopes, 44, 49, 54
Slurry pond, 91—114, 118, see also Coal slurry
SMCRA, see Surface Mining Control and Reclamation Act
Smithsonian Institute, 17
Snags, 35
Sodicity, 43, 54
Sodium absorption ratio (SAR), 43, 51—53
Soil, 28
   alkalinity, 42
   atmosphere, 43—44
   bulk density, 44, 52
   chemical properties of, 28—29
   clean-up, 72
   fertility, 42—43, 49
   fertilization, 42—43
   matrix, 43
   moisture, 41—42, 48—49, 92
   nutrients, see Nutrients
   pH, 42
   porosity, 44
   salinity, 43
   sodicity, 43
   temperature, 44
   texture, 44
   toxicity, 42—43
   treatment
      after excavation, 71—74
      chemical treatment, 73
      clean-up, 72
      fusion/vitrification, 76
      greater depths, 75
      heat treatment, 74
      in situ, 75—76, 83

      microbial treatment, 73—74
      mobile installations, 74
      options, 72
      physical separation, 73
      solidification, 75—76
      solvent extraction, 73
      stabilization/solidification, 74
      surface treatment, 75
   type, 25
Solidification, 75—76
Solvent extraction of contaminants, 73
Source areas, 33
Spawning habitat, 33
Species, 18, 19, 125—126, 164—165
Spoils, 44, 47, 145, 150, 152, 156
Stabilization/solidification of soil, 31, 74, 83
Stable ecosystem dynamics, 33
Stockpiling and respreading of soil, 16
Straight ditch channelization, 25
Streams, see River and stream restoration
Strip pits, 115—122
Structure, 3, 6
Substitution of alternative qualities or characteristics, 3
Substrate, 31
Succession, 17, 18, 121
Sulfuric acid, 92
Sun Spot Mine project, 100
Superfund, 62
Surface cover, 44
Surface-mined areas, 40, 116, 139—162, see also Abandoned mines
   approximate original contour, 140—141, 143, 150, see also Approximate original contour
   bond release, 140, 147, 150, 151
   central Appalachian mining region, 143—144
   compaction, 146
   controlled overburned placement, 143—147
   geologic materials as topsoil substitutes, 144—145
   highwall-bench-outslope topography, 140—141
   institutional constraints to improved reclamation, 158—160
   mine soil construction, 145—147
   natural topsoils vs. topsoil substitutes, 143—144
   N-fixation, 149—150
   performance standards, 140
   P-fixation, 148—150
   postmining landforms, 150—158, see also Postmining landforms
   Powell River project, 140, 161
   reclamation success, 143
   regulations, 119, 160
   rehabilitation, 143
   revegetation, 140, 147—150
   spoil handling, 156
   topsoil substitutes, 143
   treatments, 148
   vegetation, 147—150
Surface Mining Control and Reclamation Act (SMCRA), 140—141, 143, 147, 150—154, 159—160

Surface treatment, 75, 148
Symposium on restoration ecology, 14
Synthetic ecology, 13—14
Synthetic experiments, 19
Synthetic membranes, 81
Systems approach, 27

T

Taboo ecosystems, 167
Targets at risk, 64—65
Taxonomic similarity index, 31
Temperature, 44—45
Test of ecological ideas, 15
Tool for discrimination, 18
Topography, 44
Topsoil substitutes, 143—145
Toxicity, 42—43, 65
Traditional taxonomic schemes, weakness of, 18
Transfer of a pollutant, 28
Transplantation, 29, 167—169
Trash catchers, 34
Tree plantings, 94—95
Tree retards, 35

U

Underpass deflectors, 34
University of Michigan Biological Station, 3—5
University of Wisconsin-Madison Arboretum, 14
Upstream migration, 30
Urban ecosystem, 65—66

V

Vegetation, 48, 51, 53
    abandoned mines, 46—47, 52
    coal slurry, 95
    covering system, 82
    establishment of, 147—148
    monitoring metal levels in, 84
    surface mining, 147—150
Velocity, 31
Visual inspection, 84

W

Waste materials, 79—80
Water, 27—28, 40—43, 49, 54, 118—119, 171—172

West Germany ecosystems restoration, 163—186
    biomass removal, 173—175
    case studies, 176—181
    creek ecosystem, 177, 180—181
    dynamic approach, 166
    ecological qualities restored, 183
    ecosystems which cannot be restored, 166—167
    federal conservation law, 166
    Fliede-Fulda marsh, 176—179
    flooded gravel pits for lake development, 172—173
    implantation, 169—170
    legal framework, 165—166
    losses, 164—165
    natural colonization, 170—171
    other methods, 175—176
    partial transplantation, 169—170
    social background, 165—166
    static approach, 166
    summary of examples, 183
    taboo ecosystems, 167
    transplantation, 167—169
    water regime regulation, 171—172
Wetlands, 7
    aerial inventory methods, 120—121
    artificial, 8
    biological elements, 118—120
    coastal, 124
    creation in coal slurry ponds, 91—114, see also
        Coal slurry
    fisheries, 118
    hydrological plans, problems with, 135
    management tactics, 118—120
    natural, 7
    natural succession, 116
    plants, 93—95
    ponds, 115—122
    research, 130
    restoration to, 9, 116
    strip pits, 115—122
    succession, 121
    wildlife, 118
Wildlife, 96—97, 116, 118
Within-substrate migration, 30
Worker safety, 85

Z

Zn, 47, 49, 52—53